Praise for Stephen Budiansky's

Blackett's War

Recommended Reading, *Scientific American*

"Budiansky has mastered the difficulties of the story, making it very read-able and compelling . . . an important work."
—*New York Journal of Books*

"A wonderful revisionist history of how intelligence derived from Bletch-ley Park's breakthroughs combined with Blackett's operational research to bypass and destroy the Nazi Wolfpacks." —*Fortune*

"A fascinating and skillful blend of naval warfare, science, and British social history with a richly diverse cast of characters." —*World War II Magazine*

"Little-known story of the Allied scientists whose unconventional thinking helped thwart the Nazi U-boats in World War II. . . . An excellent, well-researched account. . . . An engrossing work rich in insights and anec-dotes." —*Kirkus Reviews* (starred review)

"The little known history of a linchpin in the Allies' victory over the Nazis: Patrick Blackett. . . . For military history and science fans alike."
—*Publishers Weekly*

Stephen Budiansky

Blackett's War

Stephen Budiansky is a journalist and military historian whose writings frequently appear in *The New York Times*, *The Washington Post*, and *The Atlantic*. His previous books include *Perilous Fight*, *The Bloody Shirt*, *Her Majesty's Spymaster*, *Air Power*, and *Battle of Wits*.

www.budiansky.com

Also by Stephen Budiansky

History

Perilous Fight
The Bloody Shirt
Her Majesty's Spymaster
Air Power
Battle of Wits

Natural History

The Character of Cats
The Truth About Dogs
The Nature of Horses
If a Lion Could Talk
Nature's Keepers
The Covenant of the Wild

Fiction

Murder, by the Book

For Children

The World According to Horses

Blackett's War

Blackett's War

The Men Who Defeated the Nazi U-Boats
and Brought Science to the Art of Warfare

Stephen Budiansky

Vintage Books
A Division of Random House LLC
New York

FIRST VINTAGE BOOKS EDITION, NOVEMBER 2013

The Library of Congress has cataloged the Knopf edition as follows:
Budiansky, Stephen.
Blackett's war : the men who defeated the Nazi u-boats and brought science to the
art of warfare / by Stephen Budiansky.—1st ed.
p. cm.
Includes bibliographical references and index.
1. World War, 1939–1945—Radar. 2. World War, 1939–1945—Naval operations—
Submarine. 3. Anti-submarine warfare—History—20th century. 4. World War,
1939–1945—Atlantic Ocean. 5. Blackett, P. M. S. (Patrick Maynard Stuart),
Baron Blackett, 1897–1974. I. Title.
D810.R33B79 2013
940.54'516—dc23 2012025272

Vintage ISBN: 978-0-307-74363-3

Author photograph © Martha Polkey
Front cover photograph: U.S. Navy radar scope, World War II.
National Archives and Records Administration/Photo Researchers Inc.
Cover design by Joe Montgomery
Maps and diagrams by Dave Merrill

www.vintagebooks.com

Printed in the United States of America
10 9 8 7 6 5 4 3 2 1

To Ralph Erskine

Contents

Preface

FROM 1941 TO 1943, a small group of British and American scientists, almost entirely without military experience or knowledge, revolutionized the way wars are run and won.

Applying the basic tools of their trade—a thoroughly scientific mind-set backed by little more than simple mathematics and probability theory—they repeatedly demonstrated to disbelieving admirals and generals ways to double or triple the effectiveness of the faltering Allied campaign against the German U-boats. In the grim fight for control of the Atlantic during those years of uncertainty, the scientists' unconventional insights achieved the near-miraculous in a battle crucial to the larger struggle to defeat Hitler's Germany.

The scientists who beat the U-boats never numbered more than a hundred in all, a fraction of the thousands who worked to achieve the two far better known triumphs of science in the war, the breaking of the German Enigma cipher and the making of the atomic bomb. Yet they were a collection of scientific talent the likes of which probably has never been seen before or since, certainly the oddest such collection ever assembled in one place: among them were physicists, chemists, botanists, physiologists, geneticists, insurance actuaries, economists, mathematicians, and astronomers. Six would win the Nobel Prize, in physics, chemistry, or medicine. Most were far to the left in their politics: some of the best were out-and-out Marxists, and more than a few had been committed pacifists who had come to see the defeat of the Nazis as a cause that overrode their abhorrence of war. Many were almost caricatures of the sort of unmilitary, awkward, overly intellectual civilians that military men routinely viewed with undisguised contempt.

That they were there when they were so desperately needed was the extraordinary result of a confluence of events and circumstances that I have set out to describe in the following pages: the onrush of devastating reality after decades of complacency toward the submarine menace, a political awakening of scientists brought about by the Great Depression and the rise of fascism, struggles within the militaries of Britain and the United States that pitted tradition against technical innovation and social change, and the appearance in the right place of a few unconventional political and military leaders who respected science—and of a few phenomenally accomplished scientists of great moral courage and unshakable intellectual integrity.

Patrick Blackett, a British physicist, ex–naval officer, future Nobel winner, and ardent socialist, stood at the forefront of those scientists of penetrating insight and courage. It is no exaggeration to say that few men did more to win the war against Nazi Germany than Patrick Blackett. Certainly, few who did as much as he did have been so little remembered. Partly that is because he was a difficult, private, and inner-directed man whose political views and personality did not age well in the postwar world. Most people today—myself included—will find his uncritical admiration for Stalin's Soviet Union and his doctrinaire social Marxism painfully naive, at best. But it is worth remembering that that same naïveté was the source of an idealism that we can only wish there was more of today: whatever else, Patrick Blackett was fired by a sense of justice, righteousness, and self-sacrificing courage that drove him to serve his country, and the cause of civilization itself, at the time of their utmost need.

As director of the antisubmarine analysis effort for the Royal Air Force and Royal Navy during World War II, Blackett not only helped win that battle, and the war, but in so doing founded the new science of operational research; it has been an indispensable part of military training and planning ever since, a revolution in the application of science to the art of warfare.

It is far from clear that he or any of his colleagues from those perilous and heroic days of the scientific war against the U-boats would have the chance to make such an original contribution today. The bureaucratic machinery of war has become too vast and cumbersome to leave room for the gifted improvisation and iconoclastic thinking that Blackett and his colleagues brought to bear; today's routine incorporation of science in military affairs, which they themselves helped to bring about, has ironically sharpened the lines between military and civilian expertise; and science itself has become ever more narrow, specialized, and competitive, to the point that few scientists with the intensity to achieve discoveries worthy of a Nobel Prize have time left to think about much else.

Which is our loss, and which makes their story all the more worth telling.

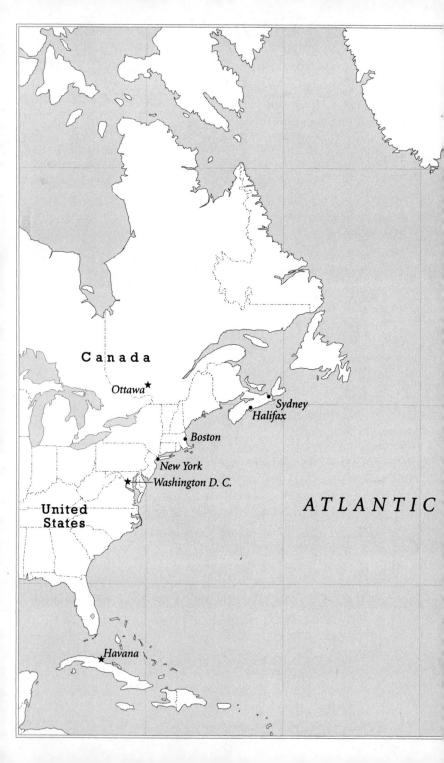

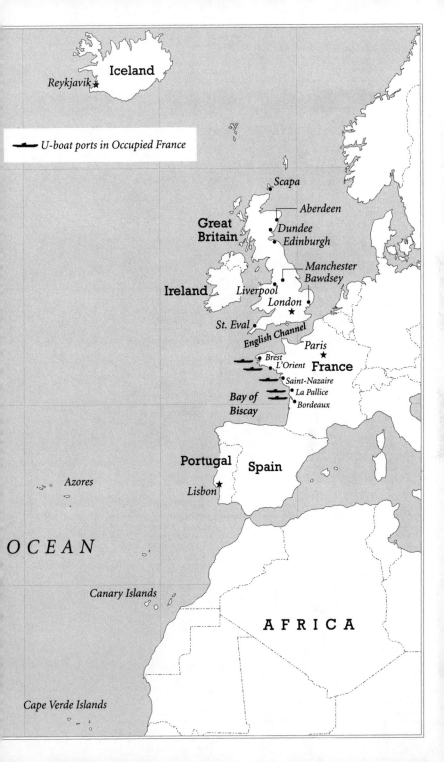

Iceland

Reykjavik ★

⬛ *U-boat ports in Occupied France*

Scapa

Aberdeen

**Great
Britain**

Dundee
Edinburgh

Manchester
Bawdsey

Ireland   Liverpool
London ★

St. Eval

English Channel   Paris ★

Brest   **France**
L'Orient

Saint-Nazaire
La Pallice
Bordeaux

*Bay of
Biscay*

Portugal   Spain

Lisbon ★

Azores

O C E A N

Canary Islands

A F R I C A

Cape Verde Islands

· Merchant ships sunk by U-boats in the Battle of the Atlantic

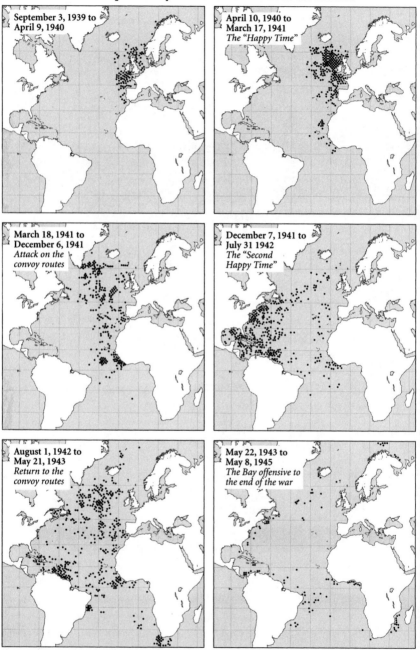

September 3, 1939 to
April 9, 1940

April 10, 1940 to
March 17, 1941
The "Happy Time"

March 18, 1941 to
December 6, 1941
*Attack on the
convoy routes*

December 7, 1941 to
July 31 1942
*The "Second
Happy Time"*

August 1, 1942 to
May 21, 1943
*Return to the
convoy routes*

May 22, 1943 to
May 8, 1945
*The Bay offensive to
the end of the war*

Chronology

1868

British engineer Robert Whitehead, working at his factory in Croatia, develops a practical self-running torpedo.

1881

American inventor John P. Holland begins trials on the Hudson River of a prototype of the first true modern submarine, dubbed by the press the *Fenian Ram*, June.

1897

Patrick Blackett is born in London, November 18.

1900

U.S. Navy takes delivery of its first submarine, the *Holland VI*, April 11.

1903

British naval college at Osborne opens as part of Admiral Jacky Fisher's scheme to expand the number of officer cadets and increase the rigor of their scientific training.

1906

German navy acquires a single submarine for evaluation, the last major naval power to do so.

1909

London Declaration, signed by all major European powers and the United States, reaffirms the rights of neutral shipping in wartime and the duty of belligerents to safeguard the passengers and crew of enemy or neutral merchant vessels taken as prizes.

1910

Blackett, age twelve, enters Osborne as a naval cadet, September.

1914

War begins in Europe, August 1.

Britain declares the entire North Sea a "war zone" and institutes a blockade of all supplies "ultimately" destined for Germany, an act not sanctioned by international law, November 2.

Blackett is a midshipman on the cruiser *Carnarvon* at the Battle of the Falkland Islands, December 8.

1915

Germany declares the waters around Britain a war zone and warns that its submarines may not be able to distinguish enemy and neutral ships, February 18.

Cunard liner *Lusitania* is sunk by a torpedo fired by a German submarine, killing 1,198 including 128 Americans, May 7.

After the sinking of two more British passenger vessels, Germany withdraws its U-boats from British waters following strong American protests, September.

1916

Blackett is at the Battle of Jutland aboard the battleship *Barham*, May 31.

1917

German U-boat force reaches 120 operational boats, January.

On the Kaiser's orders, U-boats commence an "unrestricted campaign" against British shipping, sinking merchant vessels without warning, February 1.

Sinkings by U-boats surpass 500,000 tons a month, February and March.

Citing Germany's abandonment of "all restraints of law or of humanity" in its U-boat campaign, America declares war on Germany, April 6.

After months of resistance, the British Admiralty agrees to begin convoying merchant ships, leading ultimately to a dramatic decline in sinkings by U-boats, late April.

1918

Prototype of asdic, or sonar, for detecting submerged submarines is tested aboard a British research vessel.

With its army collapsing, the German government accedes to American insistence upon an immediate halt to the U-boat war as a condition for peace negotiations, October 20.

Armistice ending the Great War, November 11.

In compliance with terms of the Armistice, 114 German U-boats surrender at Harwich, November 20–December 1.

1919

Blackett arrives at Cambridge, January 25.

1921

Blackett graduates from Cambridge with a first in physics and is awarded a research fellowship to the Cavendish Laboratory.

1922

German navy establishes a Dutch front company, IvS, to begin secretly building submarines in violation of the Versailles treaty, April.

1926

General strike in Britain, May 3–13.

1933

Nazi Party takes power as Hitler becomes chancellor of Germany, January 30.

Blackett announces at the Royal Society his discovery of the positron, for which he will later be awarded the Nobel Prize in physics, February 16.

Blackett moves to University of London, autumn.

1934

Physicists at Cavendish Laboratory sign a protest against the use of science for military purposes, June.

Professor F. A. Lindemann writes to *The Times* urging the scientific investigation of air defenses against bomber attack, August 8.

1935

Committee for the Scientific Survey of Air Defence begins work, with Henry Tizard as chairman and Blackett as a member, January 28.

The committee conducts the first experiment establishing the feasibility of detecting aircraft with radio waves, the genesis of radar, February 26.

British government announces Anglo-German Naval Treaty, abrogating restrictions of the Versailles treaty and permitting Germany to build a submarine force equal to Britain's while pledging its adherence to international prize law in submarine warfare, June 21.

1937

Blackett moves to University of Manchester.

The term "operational research" is coined by A. P. Rowe to describe the work of a scientific section at the RAF's Bawdsey Research Station studying the integration of radar into the British fighter defense system.

1938

Britain and France accede at Munich to Hitler's takeover of German-speaking areas of Czechoslovakia, September 29.

1939

German troops occupy the rest of Czechoslovakia, March 15.

Hitler denounces Anglo-German Naval Treaty, April 28.

German U-boat fleet ordered to sea, August 15.

Germany invades Poland, beginning World War II, September 1.

Winston Churchill joins British government as first lord of the Admiralty; the British passenger liner *Athenia* is torpedoed by a U-boat off Ireland the same day, September 3.

British aircraft carrier *Courageous* is torpedoed in the Bristol Channel, September 18.

Hitler discusses with his naval commanders gradually removing restrictions of international law on U-boat operations, September 23.

British battleship *Royal Oak* is torpedoed by a U-boat at Scapa Flow, October 13.

Hitler approves sinking without warning British and French merchantmen, October 16.

Karl Dönitz, the commander-in-chief of U-boats, issues standing order No. 154, instructing submarine commanders not to rescue passengers and crew from torpedoed ships, late November or early December.

1940

Following Germany's invasion of France, Churchill becomes prime minister, May 10.

Tots and Quots club agrees to produce the Penguin book *Science in War*, urging broader application of science to the war effort, June 12; Blackett likely contributes the section on operational research.

President Franklin Roosevelt approves Vannevar Bush's proposal to establish a National Defense Research Council, June 12.

France surrenders, June 21.

Just completed radar warning network along the British coast allows RAF Fighter Command to successfully fight off the Luftwaffe in the Battle of Britain, July–September.

U-boats begin operating from bases in France, July.

Equipped with more reliable torpedoes and aided by decrypted British radio reports on convoy movements, U-boats sink one million tons of shipping, July–September.

Tizard proposes to Churchill an exchange of technical information with American scientists and departs for Washington, August.

Blackett becomes scientific adviser to Anti-Aircraft Command and begins applying operational research methods to problems of radar and antiaircraft gunnery, August.

Night attacks by German bombers on British cities begin, September.

1941

Blackett becomes chief of the newly created Operational Research Section of RAF Coastal Command ("Blackett's Circus"), March.

Churchill issues Battle of the Atlantic Directive, ordering British forces to "take the offensive" against the U-boats, March 6.

The first ASV (anti–surface vessel) radar units are installed on British warships escorting Atlantic convoys, spring.

British code breakers at Bletchley Park, using captured materials from *U-110*, achieve their first sustained success deciphering German naval Enigma messages, May.

Hitler invades Russia, June 22.

Blackett's recommendation to change the camouflage color of Coastal Command aircraft doubles U-boat sightings per flying hour, summer.

Blackett is the sole dissenter on a British government panel that recommends the development of an atomic bomb by Britain; Blackett proposes discussing collaboration with an American program instead, July.

U.S. Marines land in Iceland and U.S. Navy begins escorting Atlantic convoys, July.

E. J. Williams, a member of Blackett's group, completes a study showing that the effectiveness of air attacks against U-boats can be improved by a factor of 10 through a simple adjustment in the depth setting and spacing of depth charges, September 11.

U.S. destroyer *Reuben James* is torpedoed by a U-boat off Iceland, October 20.

Bletchley Park code breakers appeal directly to Churchill to address their acute manpower and equipment shortages, October 21.

Blackett moves to the Admiralty as chief advisor on operational research (later director of Naval Operational Research), December.

Japanese attack Pearl Harbor, December 7.

1942

U-boat offensive begins against shipping along American coast, January 13.

Atlantic U-boats begin using four-rotor Enigma machines, interrupting Bletchley's ability to decipher messages, February 1.

Blackett opposes strategic bombing of Germany, based on calculations showing its ineffectiveness, and recommends shifting long-range aircraft to the Battle of the Atlantic, February.

E. J. Williams submits report demonstrating the effectiveness of employing long-range aircraft equipped with radar and Leigh Lights against U-boats transiting the Bay of Biscay, February.

Philip Morse is chosen to head U.S. Navy's new Anti-Submarine Warfare Operations Research Group (ASWORG), March.

Rodger Winn, head of the British Submarine Tracking Room, arrives in Washington to press the U.S. Navy to establish a similar centralized intelligence operation to coordinate antisubmarine operations, April.

RAF Bomber Command launches 1,000-plane attack on Cologne, May 30.

Cecil Gordon begins study of maintenance and flying schedules of Coastal Command aircraft that will triple their effective flying hours, June.

Captain Wilder D. Baker of the U.S. Navy warns "the Battle of the Atlantic is being lost," June 24.

U-boats return to the Atlantic convoy lanes, midsummer.

Dönitz delivers broadcast to German people warning that "even more difficult times lie ahead" in the U-boat war, July 27.

Dönitz issues orders positively forbidding U-boats to aid survivors, September 17.

U-boats begin to be equipped with Metox radar-warning receivers, sharply reducing effectiveness of Allied antisubmarine air patrols in the Bay of Biscay, late September.

American and British code breakers agree to begin "full collaboration" on the naval Enigma problem, October.

U.S. Eighth Air Force commences a series of bombing raids of questionable effectiveness against heavily reinforced U-boat bases in France, October 21.

Two squadrons of American B-24s, the first Allied aircraft equipped with centimeter-wave radar, depart for Britain, November 6.

Morse and his deputy William Shockley arrive in London to confer with British operational research experts, November 19.

Sinkings by U-boats reach their highest monthly total of the war, 802,000 tons, November.

Using captured documents from *U-559*, Bletchley code breakers resume reading of U-boat Enigma signals, December 14.

1943

Allied leaders meeting in Casablanca direct that the strategic bombing of Germany is the primary objective of the Allied heavy bomber force, January 21.

Blackett's staff produces a pivotal study demonstrating that ships are twice as safe from U-boat attack in large convoys as in small convoys, January 27.

Dönitz is named commander-in-chief of the German navy, January 30.

Atlantic Convoy Conference convenes in Washington and recommends reallocating 260 very-long-range aircraft to the antisubmarine campaign, March 1–12.

A temporary blackout in the reading of Enigma traffic causes two eastbound Atlantic convoys to fall prey to a devastating attack by forty U-boats, March 16.

Blackett engages in a bitter dispute with Lindemann and Air Marshal Arthur Harris over bombing priorities, March.

Centimeter-radar-equipped aircraft sink their first U-boat, March 22.

Admiral Ernest King establishes the Tenth Fleet to consolidate U.S. antisubmarine operations (and subsequently names himself commander), May 20.

Dönitz withdraws U-boats from the North Atlantic convoy lanes, May 23.

U.S. Army Air Forces agrees to turn over all antisubmarine air operations to the navy, July 9.

Air attacks on U-boats transiting the Bay of Biscay intensify and the number of operational U-boats at sea drops for the first time in the war, falling to sixty (half the number in the spring), July–August.

1944

Admiral King announces that the U-boats have been reduced from "menace" to "problem," April.

Allied troops land in France on D-Day, June 6.

1945

Dönitz succeeds Hitler as Führer, April 30.

Germany surrenders, May 8.

1948

Blackett is awarded the Nobel Prize in physics.

Blackett publishes *Fear, War, and the Bomb*, opposing the American monopoly on atomic weapons and denouncing the policies of the United States as the chief threat to world peace.

1952

Operations Research Society of America is founded and enrolls 500 members in its first year.

1965

Blackett is appointed president of the Royal Society.

1974

Blackett dies, July 13.

Blackett's War

An Unconventional Weapon

ON THE EVENING OF NOVEMBER 19, 1918, eight days after the armistice that ended the war to end all wars, a train from London pulled into the depot at Parkeston Quay, just outside the East Anglia port town of Harwich, and a mob of reporters, photographers, and newsreel cameramen spilled out onto the platform. Harwich had seen its ups and downs as a small North Sea port. In the Middle Ages the town prospered shipping bales of wool to the continent and importing French wines. In the seventeenth century, its dockyards served as an important supply and refitting base for the Royal Navy during the Dutch Wars; Samuel Pepys, the secretary to the Board of the Admiralty and keeper of the vain and ingenuous diaries that remain the most revealing account of life in Restoration England, represented the town in Parliament; and Harwich's thriving private shipyards may, or may not, have built the merchant ship *Mayflower,* which carried the Pilgrim Fathers to America.

A slow decline in the nineteenth century—the royal dockyards closed in 1829—was abruptly reversed in the 1880s when the Great Eastern Railway Company developed a large new port on reclaimed land a mile up the River Stour from the town center. The railway was rerouted to a new station from which passengers could transfer directly to ferries that took them on to Gothenburg, Hamburg, and the Hook of Holland; there were freight yards, a hotel, and rows of terraced housing for workers. With the coming of war in 1914 the Royal Navy requisitioned the entire port—quays, hotel, workshops, and all—and a force of destroyers and light cruisers and the 8th and 9th Submarine Flotillas moved in to guard the northern approaches to the English Channel.

And so Harwich, with its men who knew submarines and its facilities for handling them and its proximity to Germany's North Sea naval bases, was chosen as the place where an unprecedented event in the history of naval warfare was to take place on the morning of November 20, 1918. The terms of capitulation the German government had agreed to were extraordinary and humiliating, a measure of the desperation that the swift collapse of Germany's military situation had left her leaders facing. Fourteen articles of the Armistice dealt with the German navy. In addition to disarming all her warships and agreeing to have 10 battleships, 6 battle cruisers, 8 light cruisers, and 50 destroyers "of the most modern type" interned in neutral or Allied ports, Germany was to surrender outright "all submarines at present in existence . . . with armament and equipment complete."

Article 22 continued:

> Those that cannot put to sea shall be deprived of armament and equipment and shall remain under the supervision of the Allies and the United States. Submarines ready to put to sea shall be prepared to leave German ports immediately on receipt of wireless order to sail to the port of surrender, the remainder to follow as early as possible. The conditions of this Article shall be completed within fourteen days of the signing of the Armistice.

Along with the horde of reporters, British submarine officers and men had been summoned from every port to be on hand to take charge of the enemy boats as they arrived. Accommodations at Parkeston, which included three moored depot ships, were packed far beyond capacity that evening of the 19th. The one "lady reporter" in the group was chivalrously offered the hotel billiard table as a bed for the night.[1]

A heavy fog shrouded the harbor the next morning as the destroyers *Melampus* and *Firedrake*, carrying the boarding parties and their attendant pack of press hounds, got under way at 7 a.m. heading for the point where the surrender was to take place; it was the southern end of the shipping channel known as the Sledway, about eight miles east-northeast of Harwich. A British airship droned out of the mist and passed to the north, quickly vanishing again in the fog. Then a few minutes before 10 a.m. a British light

cruiser suddenly came into sight in the distance, then two German transports flanked by more British warships.

And then there they were: a line of unmistakable, long thin hulls breaking the dark surface of the water, topped by domed conning towers, proceeding in straggling order. Two airships and three flying boats kept a
continuous watch over the procession, passing and repassing low over the
enemy boats as they came on slowly toward the rendezvous point. Lieutenant Stephen King-Hall, a British submarine torpedo officer, groped to
find words to capture the incredulity he felt as he witnessed the scene from
aboard the *Firedrake*: the dangerous and reclusive predator he and his com-
rades had hunted and feared and loathed, now meekly chivvied along like
a few tame sheep. "Try and imagine what you would feel like," he wrote, "if
you were told to go to Piccadilly at 10 a.m. and see twenty man-eating tigers
walk up from Hyde Park Corner and lie down in front of the Ritz to let you
cut their tails off and put their leads on—and it was really so."[2]

A signal was given to the transports to anchor, and one by one the
line of twenty U-boats joined them under the guns of the British destroyers. Motor launches came alongside the *Firedrake* and the *Melampus,* and
the British boarding crews, two or three officers and fifteen men for each
U-boat, scrambled aboard. Not sure what to expect, the officers all carried
sidearms. "We were prepared for any eventuality except that which actually
took place," recalled King-Hall. "We were not prepared to find the Huns
behaving for once as gentlemen."

King-Hall's boarding of *U-90* went by the book, with punctilious cor-
rectness. The German officers saluted; the salutes were duly returned; the
German captain presented the signed terms of surrender—all equipment
intact and in working order, all ballast tanks blown, torpedoes on board
but disarmed, no booby traps—and the submarine's officers seemed almost
pathetically eager to be helpful, offering explanations of the operation of
the boat and its gear. The same scene was being repeated all along the line.
"My Hun," remarked one of the British officers back in Harwich that evening, "might have been trying to sell me the boat, the blighter tried to be so
obliging."

As the submarines raised anchor the British crews ran up the white flag
for the final transit into port. A strict order had been issued by the port
commander that there would be no cheering or other demonstrations, and
as the captured U-boats passed the ships in the harbor, crowded with spec-

tators, they were greeted by silence. By 4 p.m. they were moored to buoys at the head of the harbor (at what "the reporters now say we call 'U-boat Avenue,'" King-Hall sarcastically noted); a motor launch came alongside and the Germans, who had meanwhile changed into civilian garb that made them look more like peacetime caricatures of German tourists, green felt hat and all, than officers of a fierce and proud militarist state, were told to gather their belongings and get aboard. The launch took them to one of the British destroyers, which delivered them to the German transports for the trip back home, without ever having set foot on British soil.

Over the next eleven days the scene was repeated in daily succession as ninety-four more U-boats surrendered at Harwich, all without incident. Some of the German sailors inquired pathetically of the boarding crews if they thought they might be able to find work as merchant seamen in China or Japan, if Germans were now unwelcome anywhere closer to home. Two refused to return to Germany and insisted on staying in England, where they hoped to find "work and good food." Many of the surrendering boats were commanded by junior and plainly nervous young officers, their regular captains apparently having refused to make the humiliating voyage; others flew the red flag of the revolutionaries who had seized parts of the German fleet in the waning days of the war, their captains elected by the crews and holding commissions signed by the Sailors' and Soldiers' Committee; in other boats the crews sullenly refused to obey orders of their regular officers except when it was clear that the order would be backed up by the British officer on board. Most, especially the older men who were members of the naval reserve and had been merchant sailors before the war, seemed simply relieved that it was over, and bade farewell to their boats with dry eyes and no apparent regrets.[3]

In December 1918 the Allied Naval Commission discovered 62 additional seaworthy U-boats and another 149 still under construction at German bases and yards and ordered the immediate surrender of any that could sail or be towed and the destruction of the rest. The German government was warned that failure to turn over all of its U-boats intact would be answered by the Allies with the permanent occupation of its island naval base of Heligoland. The captured fleet, 176 boats in the end, was parceled out among the victors, most going to Britain and France, with token specimens awarded to the other Allies; Italy received 10, Japan 7, the United States 6, Belgium 2.[4]

One of the behemoths of the German U-boat fleet—the "super-submarine" *Deutschland*, originally constructed as a blockade-runner with a cargo capacity of 750 tons—was scheduled to be broken up. But at the urging of members of Parliament it was instead towed to the Thames in October 1919 and exhibited to raise money for the King's Fund for Sailors. "Poetic justice," one member of Parliament declared with satisfaction.[5]

But the real satisfaction to those who had battled this new undersea menace had come four months earlier, with the German signing of the Treaty of Versailles on June 28, 1919. Among its hundreds of detailed military stipulations, specifying everything from the maximum number of officers permitted in the headquarters of a cavalry division (15) to the number of rounds of ammunition that could be stocked per rifle or carbine (400), was Article 191, which declared: "The construction or acquisition of any submarine, even for commercial purposes, shall be forbidden in Germany." In the course of the war Germany's submarines had sunk over 5,000 Allied merchant vessels, totaling 12 million tons of shipping. No one before the war had imagined that the submarine, barely more than an inventor's crackpot dream a few years earlier, would have been capable of bringing Great Britain, and its mightiest fleet on earth, to the edge of catastrophe in 1917; no one had imagined the inhumanity that the submarine would make routine and inevitable once it was unleashed in the only way it could be truly effective as a weapon of total war. But what man's sordid ingenuity for destruction had created, the majesty of international law as decently dreamed of by Woodrow Wilson, and cynically seconded by his more worldly wise allies in London and Paris, would nobly or ignobly contain.

IT WAS THE SPEED of technological progress, more than anything, that had left conventional naval thinking so far behind. In 1887, a dredger working its way through Kiel harbor struck a large obstacle buried in the mud. After considerable effort the object was grappled to the surface, and when layers of seaweed and encrusted barnacles were scraped away, the astonished workers found themselves staring at a 40-ton, 26-foot-long hulk of sheet-iron plates, shaped like a child's drawing of a rather rectangular whale. The relic turned out to be *Le Plongeur Marin*, a three-man submersible conceived by a Bavarian artillery corporal named William Bauer, who had dreamed of a daring strike against the blockading enemy fleet during Germany's war with Denmark in 1850. During a trial dive on February 1, 1851, a slug of iron ballast that could be screwed forward or aft on a threaded rod to control the boat's inclination came loose and slid to the bow. The *Plongeur*, true to its name, promptly nosed straight down and headed for the bottom. Bauer and his crew of two came to an abrupt halt sixty feet below the harbor's surface. With remarkable presence of mind, Bauer realized their only hope was to flood the compartment with water to compress the remaining air trapped inside until its pressure equaled the force of the seawater bearing against the hatches on the outside, allowing the hatches to swing open. Convincing his two sailors to carry out these orders was another matter, but after a violent argument—and five hours of effort—his desperate plan succeeded, and the men "shot up to the surface like corks out of a champagne bottle," one witness recounted.[6]

The inventor of *Le Plongeur Marin* was scarcely distinguishable from scores of other obsessives, monomaniacs, and crackpots who for centuries had been drawn to the idea of undersea travel; only the dream of flying through the air held a greater allure for the legion of half-mad inventors. During the American Revolution and the War of 1812, intrepid American experimenters had built tiny submersibles with which they attempted to fasten explosive charges to British warships on dark nights off New York and Long Island. They all failed, but rattled the British nonetheless, who were left sputtering with indignation over such unsportsmanlike conduct; beside submarines the Americans had sent floating kegs of explosives adrift toward the anchored enemy ships and planted booby traps in decoy vessels filled with naval stores that they let fall into British hands. "It appears the Enemy are disposed to make use of every unfair and Cowardly mode of Warfare," fumed the British admiral commanding the Royal Navy's forces fighting America in the War of 1812.[7]

There was always something of the ludicrous and harebrained, or at least tragicomic, in these inventors' schemes: their makeshift submersibles, looking about as seaworthy as a whiskey barrel, leaked furiously and were barely controllable. All were conceived upon the extremely dubious operational premise that they would approach an enemy ship undetected, laboriously augering a hole or plunging a harpoon into its wooden hull below the waterline to affix a charge, then retreat a safe distance, again undetected, before finally yanking a lanyard to set off the explosion. In everything from their basic sailing properties to their purported military utility they depended as much on the sheer daredeviltry of their pilots as on the technological ingenuity that went into their conception. Their hand-powered screws and cranks and valves looked more like something out of Leonardo da Vinci's sketchbooks of the fifteenth century than the naval architecture of the nineteenth.

The first of these hand-cranked, barely submersibles to carry out a successful attack on an enemy warship was the Confederate *H. L. Hunley* in the American Civil War, but at a terrible cost to her own crewmen. Thirteen men were killed in initial training mishaps, the men drowning or suffocating when the boat twice sank to the bottom of Charleston harbor. Shortly after her third set of volunteers managed to attach and detonate an explosive mine that sent the Union steamship *Housatonic* to the bottom on February 17, 1864, the *Hunley* sank for the third and last time, perhaps when the oxygen inside the crew compartment was exhausted. All eight men on board died.

But in the thirty-six years that Bauer's *Plongeur* lay in the mud beneath Kiel harbor, everything changed: the modern submarine was born. Much of the credit went to another man whom it would have been easy to mistake for just another crank. John P. Holland was an Irish schoolteacher and music instructor with a high school education, poor eyesight, frail health, and improbable dreams of building both a flying machine and a submarine. In 1873, aged thirty-two, he took passage by steerage to America. His two brothers, who had joined the Irish independence movement, had preceded him, fleeing the British authorities. In February 1875, now teaching at a parochial school in Paterson, New Jersey, Holland sent the U.S. Navy Department a plan for a one-man, pedal-powered submarine. The navy dismissed his design as impractical.

If the navy wasn't interested, it was exactly the kind of idea to appeal to the militant fringe of Irish Americans who since the 1860s had been hatching increasingly wild and daring schemes to strike British interests around the world in the cause of Irish independence. A few hundred members of the Fenian Brotherhood, a secret society founded in New York City in 1857, had launched a comic-opera "invasion" of British Canada at the end of the American Civil War with the idea of holding the territory hostage in exchange for Irish freedom; the U.S. government in the end paid their train fare home in return for their promise to invade no more foreign countries from American soil. More serious was the raid carried out by the Fenians on the British penal colony at Freemantle, Australia, in the summer of 1876; the arrival in New York on August 19, 1876, of six Irish political prisoners freed in that coup—the men had all been sentenced to penal servitude for life—made headlines around the world.[8]

Over the next four years the Clan na Gael, the organization that had since largely succeeded the Fenian Brotherhood, collected more than $90,000 in contributions to its "Skirmishing Fund," intended to support a campaign of terror against the British. It was at this time that Holland, probably through his militant brother Michael, made contact with the fund's trustees. In early 1877 at Coney Island he demonstrated a small working model of a powered submarine. The Fenians agreed to support his experiments, and just over a year later, on June 6, 1878, the 14-foot-long *Holland No. 1*, its balky engine temporarily powered by a jury-rigged steam line run from an accompanying launch, set out from its dock on the Passaic River, ran along the surface at 3½ knots, dove to 12 feet, then dove again and stayed down an hour before resurfacing. The impressed Irish patriots promptly offered Holland a further $20,000 to construct a full-size version that might be capable of striking an actual blow against British tyranny.

Despite efforts to keep the construction of the new boat a secret, word of Holland's experiments spread quickly as soon as he began test runs in the Morris Canal Basin, on the New Jersey side of the Hudson River, in June 1881. A reporter from the *New York Sun* showed up soon thereafter, tried and failed to persuade Holland to give him a story about the submarine, then went ahead and wrote his story the next day anyway. Among other things he inventively asserted that the vessel was called the *Fenian Ram*. The name, woefully inaccurate but unimpeachable as publicity, stuck.

The whole enterprise was ludicrous: a self-taught inventor, shadowy revolutionaries, a secret war chest, incoherent conspiracies. Except that Holland had, quite simply, invented the modern submarine there on the shores of the Hudson. His new boat was powered by a 17-horsepower petroleum engine that drove the screw as well as two compressors that supplied compressed air to keep the balance tanks trimmed; by partially blowing out the water in the tanks, small changes in the boat's weight as fuel was expended and projectiles fired could be compensated for. The problem of longitudinal stability—the tendency of a submarine underwater to rock back and forth along its length like a seesaw—had bedeviled all earlier inventors and would continue to plague Holland's rival designers even for years to come. Holland ingeniously solved it by maintaining a fixed center of gravity and positive buoyancy: centrally located seawater ballast tanks together with compressed air reservoirs in the bow and stern stabilized the boat. Instead of relying on ballast to make the boat sink on an even keel like a lead weight (or, worse, employing awkward and not very effective contrivances like vertically projecting screws to propel the boat up and down), Holland's design used the dynamic force of the boat's forward motion, acting on diving planes at the stern, to drive it underwater even while preserving a small positive buoyancy. Being able to dive and surface "like a porpoise" in this fashion, Holland explained, allowed for quick dives to evade an enemy and also kept the boat maneuverable underwater, instead of wallowing like a waterlogged drum. It was the principle of all successful submarines since.

The boat was equipped with an air-powered gun that could fire a 100-pound charge of dynamite 50 yards underwater or 300 yards through the air. The *Fenian Ram* made 9 miles an hour on the surface, and probably almost as much submerged with its engine breathing a supply of stored compressed air. "There is scarcely anything required of a good submarine boat that this one did not do well enough, or fairly well," Holland later wrote.[9] It was not an idle boast: in its essentials—propulsion, balance, weaponry—the *Fenian Ram* had all the working ingredients of a true submarine.

Meanwhile, the Fenians were falling out among themselves. There were accusations of misuse of the Skirmishing Fund and demands for accounting from some discontented members, with others objecting that so much had been spent on this speculative "salt water enterprise" at all. Fenian leader John Devoy was accused in a front-page article in a rival New York Irish newspaper of disrespecting "the intent of the donors" to the fund who were expecting more immediate and visible results. (Devoy shot back, "England always gets her dirty work done among Irishmen by ardent 'patriots' who want value for their money and ten cents worth of revolution every week, or an Englishman killed every once in a while, and the breed is with us yet.")[10] In November 1883, with a court case pending in New York that threatened to tie up the fund's assets, a few of the Fenian leaders decided to save their investment. Armed with a pass bearing Holland's forged signature, they entered the docks at the Canal Basin and towed the *Ram* off into the night to New Haven, Connecticut. There Holland's invention fell quickly into dereliction. The Fenians tried a few times to take her out under her own power but were so inept that the harbormaster declared the boat a menace to navigation and forbid any more trials. The boat was then moved up the Mill River and stored in a lumber shed at a brass foundry owned by one of the Fenians; her engine was later stripped to power a forge at the foundry. In 1916 the *Fenian Ram* was carted back to New York City and exhibited at Madison Square Garden to raise money for victims of the Irish Easter Week rebellion earlier that year, her only actual service in the cause of Irish independence.[11]

Holland was furious over the theft of the boat ("I'll let her rot on their hands," he declared) but the experience he had gained from $60,000 worth of the Fenians' support, as well as the ensuing publicity he had garnered, at last attracted the U.S. Navy's attention. Lieutenant William W. Kimball, who had followed press reports of Holland's work, came to New York, took him to dinner, and listened raptly as Holland explained the principles of stability, dynamic diving, and maneuverability behind his design. Kimball was promptly sold, and spent the next two decades wrangling with the navy bureaucracy and Washington politics as he tried to get the navy to acquire its first submarine. "Give me six Holland submarine boats, the officers and crews to be selected by me, and I will pledge my life to stand off the entire British flying squadron ten miles from Sandy Hook without any aid from our fleet," Kimball declared at one hearing before the Senate Committee on Naval Affairs in 1896. Captain Alfred Thayer Mahan, the grand old man of sea power strategy, added his support in a letter to the committee chairman, stating that in his view the submarine boat would be "a decisive factor in defending our coasts" in any future conflict.[12]

Holland added two final refinements to the design that the U.S. Navy would eventually acquire. First, an auxiliary electric motor, powered by a battery of sixty wet cells, now provided the power to propel the submarine while submerged. The batteries could be recharged by clutching the electric motor to run backward, thus acting as a generator, when the boat was surfaced and its drive shaft was being turned by the boat's 45-horsepower gasoline engine.

The other new feature was a single torpedo tube that could fire a self-running, or "locomotive," torpedo. The modern torpedo had its genesis in 1864 when a British engineer named Robert Whitehead, who was manager of a small factory in the port town of Fiume in the Austro-Hungarian Empire (now Rijeka, Croatia), was approached by an Austrian naval officer with the basic idea. By 1868 Whitehead had perfected a mechanism to keep the torpedo running at a preset depth: a pendulum automatically kept the torpedo on an even keel by adjusting a tail fin to swing up or down if the nose pitched down or up; a hydrostatic valve, actuated on one side by the external water pressure and on the other by a spring adjusted to a preset tension, was similarly linked to the fin to make the torpedo rise or fall if it deviated from a predetermined depth. A chamber of compressed air drove a motor that spun a propeller, and a pistol on the nose set off a charge of 100 pounds of guncotton on impact with a target. Early models had a range of 300 yards and traveled at 6 knots; by 1900, after the British government had bought an interest in Whitehead's invention and a series of improved designs were developed, torpedoes were routinely achieving speeds of 30 knots and ranges of 800 yards or more.[13]

Deployed at first on small fast warships—torpedo boats—the weapons caused serious concern from the late 1870s on for their potential threat to the battle fleets of the world's great navies; other fast warships known at first as torpedo boat destroyers, and later simply *destroyers*, were developed to counter them.[14] The idea of marrying the torpedo to the submarine took no great leap of imagination.

On April 11, 1900, the U.S. government issued payment of $150,000 and took official delivery of the *Holland VI*, which would become the U.S. Navy's *SS-1;* within a few months orders for seven more improved "Holland-type" boats were issued. The improved models followed Holland's basic plan but were bigger and more powerful, 64 feet long and equipped with a 180-horsepower engine. Over the next six years Holland's Electric Boat Company licensed foreign rights to Britain, Russia, the Netherlands, and Japan. The first submarines of all the world's major navies—save only Germany's—would be built to Holland's design.

HOLLAND ONCE DISDAINFULLY REMARKED that naval officers didn't like submarines "because there is no deck to strut on." Yet even the submarine's most enthusiastic supporters acknowledged that they could not imagine it ever having more than a distinctly limited role in naval operations. Above all, the submarine would be a *defensive* weapon, they believed. In his testimony to the Senate Naval Affairs Committee, Kimball had melodramatically offered to stake his life defending New York harbor: it was no coincidence he had invoked that scenario. The submarine's primary function, in his view, would be to hold the first line of coast defenses, just beyond the range of shore guns. Submarines might also be useful, he told the committee, in harassing and driving off a blockading squadron, carrying communications through hostile lines, clearing minefields, and guarding channels and other narrow waterways against a fleet attempting to enter them.[15]

The last thing that any naval strategists contemplated for the submarine was that it would be a commerce raider. There were unshakable technical, legal, and strategic reasons for that conclusion. To cruise effectively against an enemy's trade required substantial range and endurance, qualities the submarine seemed unlikely ever to possess. Capturing enemy merchant ships in wartime was a perfectly legitimate practice under the established rules of international law, but again submarines were remarkably ill-suited to the particulars.

The rules for taking prizes at sea were grounded in precedents of admiralty courts going back centuries, more recently affirmed by a series of international conventions. The nuances of admiralty law were sufficiently complex to keep lawyers busy, but the basic principles for what constituted a legal capture, and the procedures a captor had to follow to ensure the legality of his actions, were universally recognized and uncontroversial. Warships of a belligerent power could stop, board, and search any merchant ship they encountered on the high seas; ships or goods found to be owned by enemy nationals could be claimed as prizes, as could neutral vessels transporting arms or other "contraband of war" to an enemy port. But every capture was subject to a proceeding in the capturing country's admiralty courts upon their return. Showing an often surprisingly fierce independence, admiralty courts rarely hesitated to disallow improper captures, even awarding damages to a ship's owner, when the letter of the law had not been followed.

In general, the law permitted a captor to burn or sink a prize and its cargo only if its enemy ownership was beyond doubt; even in that case, the captor had an unambiguous duty first to remove the crew and passengers and to preserve the ship's papers and other documents for examination by a prize court to validate his action. The London Declaration of 1909, an agreement on the laws of naval war signed by all the major European powers and the United States, reiterated all of these points, in particular the rules respecting the rights of neutrals. Although never formally ratified, the London Declaration for the most part was simply a restatement of the existing admiralty court precedents that constituted the international law of naval war. Notably, while offering one exception to the rule that ships of neutral countries always had to be taken into port for "the determination of all questions concerning the validity of the capture"—Articles 49 and 50 permitted the destruction of neutral vessels that were otherwise subject to capture and condemnation in the case "of exceptional necessity," if "the safety of the warship" or "the success of the operations in which she is engaged at the time" would be endangered—the declaration underscored the one inviolable principle: "Before the vessel is destroyed all persons on board must be placed in safety."

Submarines could hardly take on board "all persons" from a freighter or passenger liner before sinking her with gunfire or torpedoes; nor did they have men enough of their own to spare to place aboard a captured vessel as a prize crew; nor the armament to escort a prize to port and protect it from recapture along the way by an enemy warship.

But it was the grand strategic ideas of navy men of the early twentieth century that spoke loudest in relegating the submarine to a small, and largely defensive, supporting role. Mahan's writings on sea power theory had been enormously influential in all major naval powers of the world, and if there was one sacred truth in the gospel according to Mahan, it was that navies existed to defeat an enemy's navy. Concentrating one's naval power in a single mighty fleet would force an enemy into a climactic fleet-on-fleet battle that would decisively secure control of the oceans for subsequent operations; meanwhile, the very threat posed by such a concentrated force would compel the enemy to concentrate his forces as well to avoid defeat in detail, thereby leaving his own coasts vulnerable and checking his ability to conduct smaller forays. Commerce raiding was, in this view, a fatal dispersion of effort, or at best an inglorious sideshow to the real action.

Certainly that was Germany's conception in the years leading up to 1914. A British diplomat who visited Berlin before the war seeking a diplomatic end to the spiraling naval arms race between the two powers found the Kaiser a devotee of Mahan. Wilhelm II had read the American captain's book *The Influence of Sea Power Upon History* and was convinced that German greatness depended on control of the seas. "Germany," the Kaiser declared in 1900, "must possess a battle fleet of such strength that even for the most powerful naval adversary a war would involve such risks as to make that Power's own supremacy doubtful." Since launching its crash naval construction program in 1898, Germany had set more and more ambitious targets as it sought to challenge Britain's supremacy, eventually reaching a mighty seagoing force of sixty battleships and battle cruisers. The German High Seas Fleet was conceived on pure Mahanian lines: a series of battle squadrons each built around a core of seven or eight of the huge ships attended by flotillas of light cruisers and destroyers. Winston Churchill, shortly after taking office in 1911 as first lord of the Admiralty—the top civilian minister of the navy—warned that the Germans were building a sea force designed for "attack and for fleet action," not merely for the defense of Germany's overseas colonies and trade. The Germans, he concluded, were preparing "for a great trial of strength."[16]

In 1901 Admiral Alfred von Tirpitz, the master architect of the rising imperial fleet, had bluffly declared, "Germany has no need for submarines."[17] It was only in 1906 that the German navy acquired a single submarine for evaluation purposes, the last of the major naval powers to do so. *U-1*—the "U" stood for *Unterseeboot, U-boot* for short—was workable if unremarkable in its design. The first German submarines were powered by noisy two-stroke kerosene engines that belched huge plumes of thick white smoke and showers of embers, making them visible from afar day or night. Tall vent stacks mounted on the deck had to be detached and stowed before the boat could dive.

By 1910 German designers and shipyards had started to catch up with their rivals. From *U-19* on, German submarines were equipped with quieter, safer, and more reliable diesel engines. These new boats were also substantially larger, oceangoing vessels, 210 feet long, with four torpedo tubes (two fore and two aft), carrying 100 tons of fuel, giving them a theoretical range of as much as 10,000 miles. But only ten of the diesel-electric U-boats had been completed when war began on August 1, 1914. Overall, Germany's submarine fleet remained minuscule, numbering just twenty-four, a third as many as Great Britain's.[18]

Convinced that Britain's Grand Fleet would immediately descend upon the Heligoland Bight for the climactic confrontation foreseen by Mahan's theories, the German naval staff ordered all of its U-boats into a static defensive line with the start of the war. The U-boats were kept on the surface, moored to buoys, as part of a multilayered defense around the German naval base at Heligoland Island. The plan was for an outer ring of destroyers to launch torpedoes and then fall back, leading the enemy through the line of U-boats, which would submerge and launch their weapons; an inner ring of torpedo boats would then try to further harry and whittle down the enemy before the High Seas Fleet sallied forth to join the battle in earnest.[19]

But the British confounded expectations by holding back: in place of the grand climactic battle came anticlimactic stalemate. As early as October 1914 the commander of the German submarine force, Fregattenkapitän Hermann Bauer, chafing at the absurdity of keeping the bulk of his boats tied up to buoys in a circle around Heligoland for a battle that might never come, was urging the potential of the U-boats to strike a devastating blow at British oceangoing commerce. The commander of the High Seas Fleet relayed Bauer's arguments in a report to the chief of the naval staff, Admiral Hugo von Pohl: "I beg to point out that a campaign of U-boats against commercial traffic on the British coasts will strike the enemy at his weakest spot, and will make it evident to him and his allies that his power at sea is insufficient to protect his imports."[20]

Pohl demurred. He objected that such a course would plainly violate the international laws of naval war (and the German navy's own Naval Prize Regulations, which required strict adherence to accepted rules for the capture of merchant vessels); moreover, the danger of accidentally sinking neutral shipping might bring America or other neutral powers into the war against Germany. The naval staff, however, soon received astonishing proof of what the U-boats could do if given the opportunity. In a few brief cruises in and around the British Isles, a half dozen of Bauer's U-boats had quickly dispatched seven British warships: five cruisers, a submarine, and a seaplane carrier. The events caused "a decisive turn" in Germany's naval thinking, recalled Admiral Reinhard Scheer, who would succeed Pohl as commander of the High Seas Fleet a year later. "The submarine," Scheer wrote in his memoirs of the war, "from being merely a coastal defense machine, as was originally planned, became the most effective long-range weapon."[21]

Britain gave Germany's naval staff the justification it was already looking for a month later. On November 2, Britain declared the entire North Sea a "war zone" and announced its intention of interdicting all ships carrying supplies "ultimately" destined for Germany. A blockade of an enemy's port, closing it to all shipping, including that of neutral countries, is a perfectly legal exercise of belligerent powers, but the laws of naval war had always imposed stringent obligations on a blockading force to ensure that it did not simply become a legal loophole which would render meaningless the rights of neutral traders. A blockade had to be openly declared, restricted to specifically named ports or coastal waterways, and maintained with sufficient force as to block all access to those places: "paper blockades" had always been illegal because they would be little more than an excuse for seizing neutral ships wherever they were encountered on the high seas, captures that would never be permitted otherwise under admiralty law. Britain was doing exactly what the laws of naval war forbid, suspending the rights of neutrals to traverse an entire sea and, worse, threatening to seize even neutral noncontraband cargos destined for neutral ports—in Denmark, Norway, Sweden, and Holland—if, according to Britain, those goods might eventually be reexported to Germany.

The senior officers of the High Seas Fleet now added their weight behind

Bauer's proposal, arguing in a memorandum to Pohl that an unrestricted U-boat campaign against all merchant ships trading with Britain could be decisive. Sinking both enemy and neutral merchantmen without warning would not only be a necessity in such a campaign, given the nature of the U-boat, but an actual virtue as well, since shipowners would quickly be unable to find crews willing to face such an appalling menace:

> As England is trying to destroy our trade it is only fair if we retaliate by car-rying on the campaign against her trade by all possible means. Further, as England completely disregards International Law in her actions, there is not the least reason why we should exercise any restraint in our conduct of the war. We can wound England most seriously by injuring her trade. By means of the U-boat we should be able to inflict the greatest injury. We must there-fore make use of this weapon, and do so, moreover, in the way most suited to its peculiarities. The more vigorously the war is prosecuted the sooner will it come to an end, and countless human beings and treasure will be saved if the duration of the war is curtailed. Consequently a U-boat cannot spare the crews of steamers, but must send them to the bottom with their ships. The shipping world can be warned of these consequences, and it can be pointed out that ships which attempt to make British ports run the risk of being destroyed with their crews. This warning that the lives of steamers' crews will be endangered will be one good reason why all shipping trade with England should cease within a short space of time. . . . The gravity of the situation demands that we should free ourselves from all scruples which certainly no longer have justification. It is of importance too, with a view to the future, that we should make the enemy realize at once what a powerful weapon we possess in the U-boat, with which to injure their trade, and that the most unsparing use is to be made of it.[22]

In a decree published on February 4, 1915, Pohl declared the entire waters around Great Britain a war zone as of February 18, warning that "owing to the hazards of naval warfare" it might not "always be possible" to distin-guish neutral from enemy ships. In the next three months the U-boats sank a quarter million tons of shipping, including the Cunard liner *Lusitania*, torn open by a single torpedo fired from *U-20* on May 7. The death toll, 1,198 passengers and crew including 128 Americans, horrified the world and set off a wave of anti-German feeling in America, all the more so when it became clear that German public opinion was defiant, even exultant, over this triumph of German arms: German newspapers loudly justified the act, insisting that the *Lusitania* was an "armed auxiliary cruiser." The U.S. State Department delivered a formal diplomatic note to Berlin protesting "the practical impossibility of employing submarines in the destruction of commerce without disregarding those rules of fairness, reason, justice, and humanity, which all modern opinion regards as imperative."[23]

The exchange of notes went on for months, while behind the scenes Germany's civilian leadership struggled to contain the crisis. Germany's chancellor said he felt he was "sitting on top of a volcano": at any moment another young U-boat captain might at the push of a button bring America into the war. Considerably complicating his position was the fact that both the German navy and the German public overwhelmingly approved of the no-holds-barred U-boat offensive. In late August he sent the United States reassurance that henceforth Germany would scrupulously follow its prize regulations and avoid attacking neutral ships or passenger liners. In September 1915, after the volcano nearly erupted again following the torpedoing of two British passenger ships, the *Arabic* and the *Hesperian*, Germany effectively suspended the campaign altogether, withdrawing its U-boats from British waters. But to Admiral Scheer, the justice—and hard logic—of U-boat warfare, for all of its inescapable brutality, remained firmly on the other side:

> In a comparatively short space of time submarine warfare against commerce has become a form of warfare which is more than retaliation; for us it is adapted to the nature of modern warfare, and must remain a part of it.... For us Germans, submarine warfare upon commerce is a deliverance; it has put British predominance at sea in question, and it has shown to neutrals what are the consequences of yielding so weakly to British policy.... Being pressed by sheer necessity we must legalize this new weapon, or, to speak more accurately, accustom the world to it.[24]

SCHEER RIGHTLY SAW that the U-boat offensive struck a direct blow not only at Britain's naval power, but even more deeply at British complacency and tradition. For half a dozen generations the ascendancy of the Royal Navy had seemed to the British people as a simple fact, if not a law of nature—or perhaps even a birthright due the British race by virtue of its very being. For more than a century the Royal Navy had been the largest in the world. In 1889 the British government had gone even further, adopting a policy requiring Britain's navy to be larger than the next *two* largest navies in the world combined. Britannia had for so long ruled the waves that it had become simply inconceivable to think otherwise.

It was inconceivable, too, because Britain *did* depend on its navy for its very existence. Two thirds of Britain's food was imported; half the world's oceangoing commerce was carried on British merchant ships. Churchill, who served as first lord of the Admiralty until 1915, would afterward speak of the ships that were his charge with his usual rhetorical flourish, and his usual kernel of essential truth: "They were all we had. On them, as we conceived, floated the might, majesty, dominion, and power of the British Empire. All our long history built up century after century . . . all the means of livelihood and safety of our faithful, industrious, active population depended upon them."[25]

The Royal Navy was not only the bulwark of Britain's empire abroad and its security at home but an institution that had put down roots deep into British society. It was vast, ponderous, imperturbable, with a nearly unbroken history of victories at sea and an administrative bureaucracy almost as indomitable. Politicians who tried to institute reforms and modernizations came and went; the navy, with its careerism and traditions, its book-length schedules of job classifications and victualing lists, its vast empire of bakers, coopers, rope makers, and dockyards, sailed serenely on. Samuel Pepys, in the 1670s, had tried to do something about the clubby system of patronage and personal favors by which the captains of the navy doled out midshipmen's commissions to friends, nephews, and others with whom they were connected by class and obligation. He instituted qualifying examinations for promotion to lieutenant and tried to give the Admiralty the power to appoint at least a few midshipmen directly on its own authority. It took merely two centuries for Pepys's reforms to be completed: only in 1864 did the Admiralty manage at last to take full control of entry into the naval officer corps and institute a regular system for admitting and training cadets to supply the navy's future officers.

That was a tactical triumph over the old snug way of doing business, but what was really needed was a thoroughgoing revolution. The advent of steam, torpedoes, turret guns, wireless, and other technical innovations from the 1870s on made it painfully clear just how much more of a complete shake-up was required in the selection and education of new officers. Ships were becoming increasingly complex machines, but the engineering officers who actually knew how to operate them remained a separate branch from the sea officers who commanded the ships and who, with characteristic condescension, persisted in looking down on anyone tainted with practical knowledge and its associations with "trade." Officers were gentlemen; engineers were dubious on that score.

The naval arms race with Germany in the thirty years leading up to the war quintupled the British navy's budget; it also created a relentless demand for new manpower, especially officers, and especially technically proficient officers. The increase in the number of smaller ships like torpedo boats and submarines meant that the ratio of officers to men required was increasing, too: the officer corps needed to expand faster than the navy as a whole. Yet the navy's administrators seemed almost paralyzed in the face of what was obviously a growing crisis. From the 1870s to the turn of the century the manpower of the service had already doubled; during the same period its officer corps shrank from about 4,000 to 3,500.[26]

Resistance to change was exacerbated by instinctive resentment on the part of senior naval officers toward any civilian administrators who presumed to know better than they did, all the more so when the presumption was correct. Serving officers referred to the civilians in the Admiralty as "frocks." The civilians returned the compliment by referring to officers as "boneheads." (It was a senior admiral's complaint that Churchill was impugning naval "traditions" with one of his proposals for shaking things up during his time as first lord of the Admiralty which prompted the famous Churchillian retort that the traditions of the Royal Navy could be summarized in three words, "rum, sodomy, and the lash.") In 1902, in the midst of the officer shortage crisis, there burst into the corridors of power in Whitehall a man who had both the determination and the moral authority to drag the navy into the twentieth century whether it liked it or not. Admiral Sir John Arbuthnot Fisher was, as Churchill would later remark, "a veritable volcano of knowledge and inspiration." He would often be called Britain's "greatest admiral since Nelson," though he had never commanded a fleet in battle. Fisher was short and stocky, "ugly" by his own reckoning, with a sallow complexion that prompted his enemies to name him "the Yellow Peril." But he cut a brash and fearless image the British public loved, and he reveled in a take-no-prisoners manner and unconcealed contempt for those he considered mired in the past. He routinely referred to fellow officers who stood in his way as "pre-historic admirals," "mandarins," and "fossils."[27]

Fisher was serving as commander-in-chief of the Mediterranean Fleet when he received word he would be the next second naval lord, the senior admiral in charge of personnel, and he promptly threw himself into the task of completely reforming officer selection and education. Besides addressing the navy's critical shortage of officers, Fisher wanted to put an end to the stigma associated with engineering by greatly increasing the amount of scientific training that all cadets received and eliminating the invidious distinction between executive and engineering officers. Under Fisher's plan all officers would have a single career path up to the rank of lieutenant; only at that point would they specialize—as sea officers, gunnery or torpedo officers, engineers, Royal Marines, or instructors. Then, at the rank of commander, they would once again come together on a single general list. The key point was that all officers would have a basic grounding in engineering and science, and all would be equally eligible for promotion to senior commands.

The proposal horrified the old guard. But Fisher was not used to losing bureaucratic battles; he marshaled public support with well-timed leaks to sympathetic reporters, and by the end of the year the Selborne Scheme—named after Lord Selborne, the first lord of the Admiralty, though it might have more accurately been called the Fisher Scheme—was a reality. To accommodate the influx of new cadets, a new school was hastily established on the grounds and stable block of what had been Queen Victoria's favorite summer home, Osborne House on the Isle of Wight. (Her son Edward VII had presented it as a gift to the nation.) Starting at age thirteen, cadets attended school there for two years before transferring to the naval college at Dartmouth, whose buildings were undergoing an equally hasty expansion.

In accordance with Fisher's insistence, the curriculum provided an unprecedented emphasis on science, engineering, and mathematics—subjects scarcely recognized by England's public schools of the time, which held fast to the belief that a gentleman's education was properly confined to the classics, the less practical the better. (Even subjects such as modern history and English were considered less respectable than Latin and Greek.) Fisher managed to recruit a senior science master from Harrow to be Osborne's first headmaster; his successor was a mathematics master from Winchester College, another of the elite English public schools. A full third of the cadets' time was devoted to engineering, including about nine hours a week in the machine shop, where they learned lathe work, toolmaking, forging, casting, and carpentry—an even more remarkable bow to the new realities of the modern technical world. It was safe to say that from 1903 on, naval cadets received far and away the best secondary-level education in the physical sciences and mathematics available anywhere in England.

In theory, admission to Osborne was open to all boys who could meet the entrance requirements, which included passing a written examination in English history, geography, French, Latin, arithmetic, algebra, and geometry. In practice, the old barriers of class had changed little. Even with Fisher's reforms, the naval officer corps remained a vocation for the sons of gentlemen. For one thing, parents had to pay the cost of tuition for a boy's four years' cadet training, which effectively kept the bar at the middle class. For another, a rigorous interview conducted by a board of examiners, which included a senior admiral, made sure that the boys who were accepted were (in the words of a publication explaining the process) "the right sort." Among the questions routinely asked were many that a twelve-year-old boy who was not "the right sort" would be unlikely to have a clue how to answer: how much did a London cab charge to carry a piece of luggage; which animals and birds would be regarded as game in the countryside; what dish did caper sauce go with; how would you address a duchess, a king, or a bishop. The results spoke loudest: the successful candidates were nearly all the sons of army or navy officers, Church of England clergymen, landed gentry, or professional men such as lawyers, doctors, and engineers.[28] It would still in many ways be a very nineteenth-century Royal Navy that would face the very twentieth-century menace of the U-boat amid all of the technological and unchivalric horrors that the Great War would usher in.

2

Cruelty and Squalor

PATRICK BLACKETT WAS undeniably the right sort. His grandfather had been a vicar, his father a "reluctant stockbroker" whose main interests were books, fishing, and collecting butterflies. On his mother's side was a long line of minor landed gentry in Shropshire; the family had held a pocket seat in Parliament for three centuries and produced along the way the occasional navy or army officer and India tea planter.

Blackett had enjoyed a typically comfortable Edwardian upbringing in Kensington and Surrey of a boy born into the English middle class in 1897. It was an era when even a fairly modest income of £500 or £600 a year—what a bank manager or other mid-level professional would earn—bought comforts on a scale unimaginable today. A new six-room villa in London could be had for £30 a year. Low income taxes, about 5 percent, and an abundance of cheap domestic labor meant that even "ordinary" households routinely could afford a live-in cook and maid, as did the Blackett family: a cook typically earned £30 a year, a parlor maid £12—the approximate equivalent of $4,000 and $1,500, respectively, in 2010 dollars.[1]

Blackett's parents were kind but distant in the approved fashion; children were not praised or shown overt affection "lest they become conceited," as he remembered.[2] But there was, Blackett recalled, a "kindly security" to his boyhood domestic life. It was a feeling matched by the snug security of British society as a whole—at least for the middle class—at a time when

King, Country, Empire, and Church still held their accustomed places in the firmament of unchallenged Edwardian respectability.

Like many scientists-to-be, Blackett had all the indications of being a born skeptic on matters of religious belief, but it never occurred to him to rebel against ritual practices that, after all, by then had far more to do with social convention than religious conviction. "I was baptised into the Church of England, then vaccinated and finally confirmed, as was the usual order in those stately days," he would write in a slightly wry account ("The Education of an Agnostic") that appeared decades later in the humor magazine *Punch*, after he had won the Nobel Prize in physics and become one of Britain's most famous scientists. "Like many others I felt some slight disappointment that confirmation left on me no noticeable mark . . . definitely with me confirmation did not take." But, he continued:

> I regularly attended Sunday morning service and no doubt it did me a power of good by extracting me for a restful hour from the wooden hut in our garden where I spent every hour out of school making wireless sets and model aeroplanes. I found that I could turn this Sunday ritual to good effect when I discovered that the enforced repose of a sermon was excellently conducive to bright ideas as to how to mount a galena crystal or to carve the propeller of a model aeroplane.[3]

In the spring of 1910 Blackett's parents entered him as a candidate for one of Fisher's new naval cadetships. "I do not remember any strong wish to go into the navy," Blackett would later remark, "nor any marked reluctance," but his enthusiasm for airplanes stood him well in the interview:

> There were many tales of these ordeals and of the unexpected questions which might be shot at one, such as, "What was the number of your taxi?" or some other test of the applicant's powers of observation. I was lucky, for the first question I was asked was what did I know about Charles Rolls' flying machine, in which he had made the first double crossing of the Channel the day before my interview. I . . . proceeded to bore the admirals by telling them more than they wanted to know about Rolls and his machine.[4]

For the next four years he received "an excellent modern and scientific education, with a background of naval history, and the confident expectation that the naval arms race with Germany then in full swing would inevitably lead to war." Blackett graduated second in his class at Osborne and was the top cadet at Dartmouth. A half century later his former divisional officer at Dartmouth dropped him a congratulatory note on the occasion of his being awarded the Order of Merit and quoted from the notes he had made about his former cadet at the time: "Games. Does not shine. Remarks on Character. Clever, quiet, and nice. Works & does well, and should turn out well."[5]

On August 1, 1914, the cadets were abruptly told to pack their sea chests; by nine o'clock that night they were on their way to the train station and then the naval barracks at Devenport. Blackett, sixteen years old, was assigned as a midshipman to the cruiser *Carnarvon*. Delayed at the dockyards, the *Carnarvon* almost missed a rendezvous at the Cape Verde Islands with HMS *Monmouth*, which had carried Blackett and eleven of his fellow cadets from England. Two months later *Monmouth* was sunk by German gunfire from the cruisers *Scharnhorst* and *Gneisenau* at the Battle of Coronel off the coast of Chile, killing all aboard. *Carnarvon* was assigned to a quickly assembled force dispatched to hunt down the two German cruisers, and on December 8, the unsuspecting German squadron stumbled right into the powerful British force, which included two battle cruisers, at the Falkland Islands. The Germans turned and fled to sea and the British steamed out in full pursuit. *Carnarvon* was the slowest of the British ships but arrived in time to take part in the bombardment that sank the *Gneisenau* and to take her survivors on board.

The young midshipman kept a desultory diary full of complaints about the tedium of patrolling vast expanses of empty seas, the incompetence of admirals, the mediocrity of the food, along with enthusiastic observations about seabirds, photography, reading (bird books, Sherlock Holmes, Kipling—"my favourite"), and a Miss Macpherson, whom he had met in Quebec. His ship touched at Rio, Montevideo, Bermuda, Barbados, and Montreal; from Montreal he sailed home for leave in June 1915 aboard the White Star liner *Megantic*.[6]

His next berth was the new fast battleship *Barham*. Flagship of the Fifth Battle Squadron anchored at Scapa Flow, she was part of the Grand Fleet still awaiting the elusive, decisive battle with the German High Seas Fleet.

The much anticipated chance came on May 30, 1916. Room 40, the Admiralty's small code-breaking unit, intercepted a German wireless message indicating that Admiral Scheer's High Seas Fleet was preparing to sortie the next day. But a fatal misunderstanding on the part of the navy's director of operations, Rear Admiral Thomas Jackson—an old-school, anti-intellectual sea officer, Jackson "displayed supreme contempt for the work of Room 40" in the bitter recollection of one of the code breakers—led to Britain's bungling her best chance.[7] The next morning Jackson concluded on his own misinformed analysis that the Germans had delayed their sailing; Britain's main battle fleet was accordingly biding its time, still seventy miles away, when the Battle Cruiser Fleet under Vice Admiral Sir David Beatty ran directly into the Germans off Jutland, Denmark. Further hesitation, miscalculation, and tactical blunders ruled the day. The battle could easily have been a victory of crushing annihilation over the Germans but ended in a tactical draw, with heavy casualties on both sides.

The accuracy of the German gunfire, and the inadequate protection of the British battle cruisers, also came as a shock. "There seems to be something wrong with our damned ships today," Beatty reportedly exclaimed to his flag captain after two of his battle cruisers were blown to bits in the first forty minutes of the battle. "Half an hour later," Blackett would recall, "the Fifth Battle Squadron passed the spot where the *Queen Mary* had disappeared. That patch of oily water, where a dozen survivors of the crew of 1,200 were clinging to pieces of wreckage, as I saw it through the periscope of the front turret of the *Barham*, gave me a strong awareness of the danger of assuming superiority over the enemy in military technique."[8] *Barham*'s squadron, deployed with the battle cruisers, was caught silhouetted against the western sky and found it hard to spot the enemy to return fire. "It was very horrible seeing the flashes, then waiting for the salvoes to fall," Blackett recorded in his diary. Some five hundred 12-inch shells were fired at the *Barham* and the rest of the squadron: "How we survived with so few hits I have no idea."

Toward evening the battleships of the Grand Fleet at last joined them but Scheer executed a well-rehearsed withdrawal in the fading light and broke off the action. It was only then that the immediate toll came home to Blackett and his crewmates:

We were then allowed to leave our stations to get some supper. . . . Many people did not know till then that we had been hit, but one realised it terribly then. There was an extraordinary reek of T.N.T. fumes, which mixed with the smell of disinfectant and blood was awful. Nearly all the killed, some twenty-four in number, were lying laid out on the deck, and many were terribly wounded, limbs being completely blown off and nearly all burnt. . . . The Padre and Paymaster had been killed in the Forward Medical Distributing station. . . . The night was very trying, waiting closed up — trying to sleep in turns.[9]

The British had lost more than 6,000 killed; 3 battle cruisers, 3 cruisers, and 8 destroyers were sunk. German losses were 1 battleship, 1 battle cruiser, 4 light cruisers, 5 destroyers, and 2,500 men. At the cabinet's insistence Churchill issued a communiqué putting a positive face on the battle, but the confusing and incomplete reports in the newspapers left most of the British public uncertain whether Jutland had been a triumph or a disaster. In Germany the Kaiser exulted. "The spell of Trafalgar is broken," he crowed. But Scheer privately told his monarch that the battle had only proved once and for all the impossibility of taking on the British navy in a head-to-head trial of strength. "There can be no doubt that even the most successful outcome of a Fleet action in this war will not *force* England to make peace," he wrote in a confidential report to the Kaiser on July 4. "A victorious end to the war within a reasonable time can only be achieved through the defeat of British economic life—that is, by using the U-boats against British trade. In this connection, I feel it my duty to again strongly advise Your Majesty against the adoption of any half-measures."[10]

ONE MONTH AFTER the inconclusive battle off Jutland, the British Army launched the largest offensive in the history of warfare. Previous attempts to break the grinding deadlock on the Western Front had failed repeatedly, at horrific cost. Even when small advances through the German lines had been gained in frontal assaults carried out with dogged courage over the tops of the trenches and through barbed wire, machine gun fire, and poison gas, they were quickly driven back by counterattacks and artillery fire from the flanks of the German line; the small breaches were never wide enough, or held for long enough, to allow the masses of cavalry held expectantly in reserve to charge through in the decisive breakthrough to the enemy's rear that would send the Germans fleeing.

The Somme was going to be different. Learning from these previous failures, the British staff spent six months drawing up plans for a massive coordinated attack along a twenty-five-mile sector. For a week, British heavy guns fired a million and a half shells at the German trenches and fortifications. A quarter of a million of those shells were sent over in the final hour before the assault began at 7:30 a.m. on July 1, 1916, with more than 100,000 men moving methodically out of the trenches and advancing toward the enemy lines.

Staggering under sixty-six-pound packs and bunched together as they tried to get through the far-too-small gaps that had been cut in their own wire the night before, the attackers made a pathetically easy target.[11] Nearly 50,000 British soldiers were killed or seriously wounded that first day, most of the 20,000 dead probably cut down in the first hour by a hundred German machine guns that had survived the bombardment in concealed, heavily fortified emplacements, and which swept no-man's-land in remorseless waves.

Meanwhile, 120 miles to the southeast, the German high command had abandoned even the pretense of engaging in anything but slaughter for slaughter's sake. In a million-man offensive begun in February against two French fortresses that guarded the road to Paris at Verdun, the German commander declared it was his intention not to capture the forts, not to break through the French lines, not to open the way for maneuver on the gridlocked battlefield, but simply to grind the French down through sheer attrition of numbers. "The forces of France will bleed to death," declared General Erich von Falkenhayn.[12] When it was over 300,000 were dead. But France had not bled to death, and the war went on.

In the middle of 1916 the explorer Ernest Shackleton, stranded in the Antarctic with an expedition that had left England shortly after the war began, arrived at the British whaling station on the island of South Georgia after a harrowing open-boat journey and a perilous traverse across the ice- and snow-covered peaks of the island.

"Tell me, when was the war over?" was the first thing he asked the station's manager.

"The war is not over," the man replied. "Millions are being killed. Europe is mad. The world is mad."[13]

The sheer, terrible anonymity of the slaughter was brutalizing to sensibilities accustomed to more heroic ideas of military valor. H. G. Wells recalled the way the Boer War, scarcely a generation earlier, had been to the ordinary Briton "a vivid spectacle" that he could look upon like "a paying spectator at a cricket or base-ball match."[14] But the collective, industrialized brand of war that the machine gun and high-explosive shell had brought left little place for heroes. The trenches of the First World War were, in the words of the French poet Blaise Cendrars, a "troglodyte" world, where soldiers literally burrowed into the ground seeking to be as inconspicuous as possible. "War, which used to be cruel and magnificent," Winston Churchill would later write of the conflict, "has now become cruel and squalid. . . . Instead of a small number of well-trained professionals championing their country's cause with ancient weapons and a beautiful intricacy of manoeuvre, sustained at every moment by the applause of their nation, we now have entire populations, including even women and children, pitted against one another in brutish mutual extermination."[15]

A war without heroes, where death was faceless and numbered in the millions, was a war with little place left for chivalry, or lingering moral scruples. In Germany, where the grip of the British blockade was beginning to take a dreadful toll, thousands of civilians were dying of starvation and food riots broke out in thirty cities.[16] When the United States in April 1916 sent a protest over yet another deadly torpedo attack by a German U-boat on a passenger ship—she was a French cross-Channel ferry, the *Sussex* (which the Germans unpersuasively insisted had been mistaken by the captain of *UB-29* for a troop transport, though the attack occurred in broad daylight under excellent visibility)—Germany again promised to adhere to prize regulations, but added that it would feel perfectly justified in unleashing the U-boats to carry out unrestricted attacks without warning:

> In self-defense against the illegal conduct of British warfare, while fighting a bitter struggle for her national existence, Germany had to resort to the hard but effective weapon of submarine warfare. As matters stand, the German Government cannot but reiterate its regret that the sentiments of humanity which the Government of the United States extends with such fervor to the unhappy victims of submarine warfare are not extended with the same warmth of feeling to the many millions of women and children who, according to the avowed intentions of the British Government, shall be starved.[17]

The U-boat campaign held an appeal for the German populace that went beyond retaliation for their suffering. It also was one of the few fronts left in this war of "brutish mutual extermination"—the air was the other—where individual heroes could still come forth and fire the public imagination. The U-boat crews were seen as daring young champions, venturing into a dangerous new realm where they pitted their dash and wit against the enemy. Successful U-boat captains had their portraits taken by the official court photographer and were presented with the Iron Cross, or the even more coveted Pour le Mérite, by the Kaiser himself. Commanded usually by a Kapitänleutnant, the equivalent of a full lieutenant in the British or American navies, the U-boats offered a rare opportunity for an ambitious junior officer to distinguish himself.

The U-boat service was like the glamorous new air service in another way: it offered an excellent opportunity for an early death. Antisubmarine warfare was in its infancy; only in 1916 did the British navy begin to acquire depth charges and hydrophones that could pick up the sound of a U-boat's propeller underwater, and even then supplies were short and effectiveness distinctly limited. Still, U-boats continuously fell prey to accidents, mines, and the simple but deadly countermeasure of ramming by destroyers or other warships that chanced to catch one on the surface. By the end of the war almost exactly half of the approximately 370 German submarines that had ventured to sea during the war had been sunk, taking 6,000 men to their deaths.

Throughout the fall of 1916 a barrage of studies and official memoranda from the German army and navy staffs kept up a drumbeat of arguments in favor of resuming all-out war on British shipping. General Erich Ludendorff, who with Field Marshal Paul von Hindenburg now effectively headed the army high command, contemptuously dismissed the American threat as empty saber rattling; the Americans had only a small professional army and the transports that would carry their troops across the Atlantic if they sought to join the fight could be easily dispatched by the U-boat fleet. The German naval staff in late December produced a study concluding that if the U-boats could sink 600,000 tons of British shipping a month, "we can force England to make peace in five months." The unrestricted U-boat campaign was not only "the right means" to achieve victory, "it is only means to that end."[18] Anticipating that it was now only a matter of time before the U-boats would be allowed to slip their leash, the navy stepped up construction and training. By the beginning of 1917, 120 oceangoing boats were in service and young officers were being rushed through an intensive three-month course at the German navy's U-School to prepare them for their new role.

OBERLEUTNANT ZUR SEE KARL DÖNITZ was one of them. In background and upbringing, he was strikingly similar to Patrick Blackett. Dönitz was six years older, but because German naval cadets entered at age eighteen he had become a naval cadet the same year as Blackett, 1910. Dönitz's family had been small farmers, pastors, scholars, part of Prussia's rising or at least aspiring middle class; his father was employed by the Zeiss optical firm. Karl collected rocks and fossils, founded a literary society and managed to convince six of his classmates to join, played the flute, and studied art.

If there was one striking difference between the two young naval officers-to-be, it was Dönitz's burning ambition to fit in. Like the Royal Navy, the Imperial Navy took a hard look at cadets' social status and charged their parents a fee that effectively limited admission to the upper middle classes; unlike the Royal Navy the German service maintained an air of Prussian aristocratic exclusivity even as its ranks swelled with the sons of the middle class inculcated with the Second Reich's hyperpatriotism and cult of soldier-ship. Admiral Fisher had done away with distinctions between executive and specialist officers; the German navy exaggerated them, demanding almost feudal deference to shipboard commanders. Officer cadets were subjected to arduous physical tasks bordering on hazing; they were taught gentlemanly refinements such as fencing, riding, and dancing; a rigid aristocratic code of honor still condoned dueling to settle tiny perceived slights.[19]

Dönitz excelled in it all. Reports by his commanders praised his diligence, enthusiasm, and perspicacity as well as his charm, popularity with fellow officers, "very good military appearance," and social deftness. A memoir the young officer published in 1917 reveals a bright but exceedingly shallow young man, describing his part in sea battles and visits to foreign ports without a hint of self-awareness or irony. The language reads like a cross between a *Boy's Life* adventure story ("There, now our salvo lands and the foremost destroyer sustains three hits! There, now another five! Suddenly, there is only his bridge foc's'le to be seen. He has had enough!") and a third-rate travelogue ("the fairy-tale town of Istanbul"). He passed out of the U-boat course in January 1917 and was preparing to take up his first posting when, at Admiral Scheer's headquarters, the long-awaited telegram from the Kaiser arrived instructing that "the unrestricted campaign shall begin on February 1 in full force." Dönitz was assigned as a watch officer on *U-39*, commanded by Kapitänleutnant Walter Forstmann, already a legendary ace who had sunk 300,000 tons of shipping and been awarded the Pour le Mérite. Later in the year Dönitz received his first command of his own, the minelayer and attack boat *UC-25*. "I felt as mighty as a king," he said.[20]

ON APRIL 9, 1917, two commonplace-looking gentlemen in civilian clothes arrived at Liverpool on the American passenger steamer *New York* and were quickly hustled aboard a special train, which departed at once for London. It was three days after America's declaration of war, which President Wilson said his country had at last been driven to as a direct consequence of Germany's abandonment "of all restraints of law or of humanity" in its submarine campaign.

During the voyage over an alert steward had noticed that the initials embroidered on one of the men's pajamas did not match his name on the passenger list and reported him to the captain as a suspicious character. The captain had a quiet laugh; he was in on the secret that the two suspicious passengers, sailing under the names S. W. Davidson and V. J. Richardson, were in fact two American naval officers, Rear Admiral William S. Sims and his aide, Commander J. V. Babcock, traveling incognito. Sims, up until a few weeks before the president of the Naval War College in Newport, Rhode Island, had been hastily summoned to Washington in late March and, with war imminent, dispatched at once to England to establish high-level contacts with the U.S. Navy's counterparts in the British Admiralty.

Upon his arrival in London, Sims was immediately ushered in to see Admiral John Jellicoe, the first sea lord. The two admirals had known each other for years, having first crossed paths in China in 1901, and Sims found the British admiral the same calm, imperturbable, frank, and approachable man he remembered, "all courtesy, all brain," betraying none of the immense burdens his job carried. British newspapers had been full of vaguely reassuring statements suggesting that the German U-boat offensive had already proved a failure and, as Sims would observe in the coming days, "this same atmosphere of cheerful ignorance" prevailed "everywhere in London society." After a few preliminary pleasantries, the first sea lord took a paper out of his desk and handed it to Sims. "I never imagined anything so terrible," Sims would recall. It was a record of actual British and neutral tonnage lost since the unrestricted U-boat campaign had begun, and it was a disaster. Sinkings had surpassed 500,000 tons a month in February and March; they would hit 900,000 tons in April if the current rate of destruction held. Sims, astonished, said that it looked like the Germans were winning the war. Jellicoe agreed. "They will win, unless we can stop these losses—and stop them soon."

"Is there no solution?"

"Absolutely none that we can see now," Jellicoe replied.[21]

Actually there was a solution: Jellicoe himself had vetoed it. Jellicoe had arrived at the Admiralty the previous December determined to shake up the antisubmarine effort. Until then there was no single command responsible for the British response to the U-boats, and the new first sea lord moved swiftly to take charge and establish a new Anti-Submarine Division. A week later a new man arrived at 10 Downing Street equally determined to put new life into the British war effort: the venerable Liberal David Lloyd George had been chosen to take the helm of the national unity government in the face of waning confidence in Prime Minister Henry Asquith. Impressed by a cabinet paper arguing the effectiveness of convoying merchant ships as a way to protect them from submarine attack, Lloyd George pressed Jellicoe to look into the idea as the first appalling sinking statistics began to arrive in February 1917.

Convoys were not, however, what Jellicoe had in mind. He thought they were impractical and ineffective, and the Admiralty's experts concurred: "It is evident that the larger the number of ships forming the convoy, the greater the chance of a submarine being enabled to attack successfully, and the greater the difficulty of the escort in preventing such an attack." Actually, *exactly* the opposite was the case; but Jellicoe was also swayed by the fact that the merchant shippers themselves opposed the idea, and at a conference at the Admiralty on February 23, 1917, both the navy men and ten merchant sea officers summoned to discuss the matter all agreed that merchant ships could not possibly manage to keep station in a large, zigzagging convoy. There were undeniable challenges in seamanship involved, but the shippers also just disliked the delays and nuisance of waiting for a convoy to assemble and having to sail under Admiralty orders, and navy officers disliked the idea of being reduced to ferrying a bunch of tubs back and forth across the ocean.[22]

But Sims was a firm believer in convoys and he began to "emphatically" press the matter, too. Sims pointed out that "sailing vessels in groups, and escorting them by warships, is almost as old as naval warfare itself." He pointedly added that the British navy already implicitly recognized the principle of convoying when it came to protecting their own battle fleets from submarines: battleships never moved without an accompanying destroyer screen shielding them. But his clinching argument was that a limited convoy system—euphemistically termed "controlled sailings"—had already been introduced at French insistence in February for colliers supplying coal to France, and had been a clear-cut success. Eight hundred colliers a month made the journey; in escorted convoys from February to April a total of five ships had been sunk by U-boats.[23]

Lloyd George quickly discovered that the American admiral was an ally and had several lengthy discussions with Sims, then began pressing his "Lord High Admirals" (as he sarcastically referred to the Admiralty Board in his subsequent memoirs) even harder, finally announcing on April 25 that he would visit the Admiralty and personally straighten out the matter. The admirals' most consistent objection had been that with 5,000 shipping movements a week at British ports, it was simply a numerical impossibility to organize convoys and provide escorts for them all. It turned out that the figure was a ridiculous exaggeration, arrived at only by counting every coming and going of vessels of every description. The actual number of large oceangoing ships that arrived and departed each week was 300. "The blunder on which their policy was based," Lloyd George would later write of the Admiralty's resistance to convoys, "was based on an arithmetical mix-up which would not have been perpetrated by an ordinary clerk in a shipping office."[24]

The institution of convoys over the next several months was nothing short of a revolution. Karl Dönitz, in his memoirs, recalled the sudden wind taken out of the U-boat offensive, "robbed of its opportunity to become a decisive factor":

The oceans at once became bare and empty; for long periods at a time the U-boats, operating individually, would see nothing at all; and then suddenly up would loom a huge concourse of ships, thirty or fifty or more of them, surrounded by a strong escort of warships of all types. . . . The lone U-boat might well sink one or two of the ships, or even several; but that was but a poor percentage of the whole. The convoy would then steam on . . . bringing a rich cargo of foodstuffs and raw materials safely to port.[25]

By the summer of 1918, sinkings had fallen to less than 300,000 tons a month.[26]

ON SEPTEMBER 26, 1918, British, American, Canadian, Australian, New Zealand, and French troops launched a huge assault against the last German defensive position on the Western Front, a twenty-five-mile-deep system of fortifications and redoubts running through northern France known as the Hindenburg Line. German morale both at the front and at home was on the verge of total collapse. Three days later Ludendorff informed the astonished Kaiser and civilian leaders of the government—who up until then had heard nothing to make them doubt the inevitability of a German victory—that the army was defeated and Germany must seek an armistice without a moment's delay if the empire was to be saved; the Kaiser must also immediately decree a parliamentary constitution to avert a revolution at home. Several days of confusion followed as the chancellor resigned and the government was without a leader capable of making a decision. On October 2 the high command's demand that the civilian government sue for peace became a virtual ultimatum. An aide sent by Ludendorff put the situation starkly: "We cannot win the war . . . we must make up our minds to abandon further prosecution of the war as hopeless."

Ludendorff would later try to throw all of the blame for Germany's surrender upon the civilian government, writing in his memoirs that he was "unable to understand how the idea ever arose that I said that the front would break if we did not have an armistice in twenty-four hours." In fact the idea came from Ludendorff's own increasingly panicked messages, including one which stated that "every twenty-four hours that pass may make our position worse, and give the enemy a clearer view of our present weakness," with "the most disastrous consequences." When the new chancellor, Prince Max of Baden—the Kaiser's second cousin and the only liberal-minded member of the royal family—balked, suggesting that such a precipitous offer of an armistice, coming so completely out of the blue, would itself have the air of capitulation, Hindenburg and Ludendorff were adamant that no delay could be tolerated: the military situation demanded an end to the fighting at once. The next day the government drafted a note stating that Germany was prepared to accept Wilson's Fourteen Points and requesting an immediate armistice and peace negotiations. (Ludendorff would later claim that he thought the message should have had "a more manly wording.")[27]

Over the next five weeks British and Commonwealth troops won a series of smashing victories, breaking through the final Hindenburg Line defenses on October 5 and then rolling forward five to ten miles a day as they captured thousands of prisoners and hundreds of enemy guns. In later years, amid the postwar mood of disillusionment and pacifism, Field Marshal Sir Douglas Haig would be remembered to history almost entirely as the man responsible for the disaster of the Somme, thus becoming the archetype of the unimaginative general who mindlessly sent millions to their slaughter. Few historians would deign to acknowledge that he was also the architect, in those final weeks, of the Allies' victory; or that with the breakthrough had come a return not just of mobile warfare but of the kind of battlefield heroism that every generation before and since rightly venerated. The advance rolled on and on, beating off frequent German counterattacks and desperate rearguard actions. In a scene right out of medieval warfare, the New Zealand Division scaled the ramparts of the ancient walled city of Le Quesnoy on ladders to rout a sizable German force holding out inside, taking 2,500 prisoners and more than 100 guns.

Well aware it would be all but impossible to resume fighting once the guns ceased firing, President Wilson insisted that an armistice incorporate all key Allied terms; so the fighting raged on as notes were exchanged between Berlin and Washington. The continuation of the U-boat campaign in the meanwhile threatened to scuttle the negotiations altogether. In the first two weeks of October, U-boats off Ireland torpedoed two passenger ships, taking more than 800 lives and setting off a rage of indignation in Britain and America that left the Allies in no mood to grant concessions. On October 14 Wilson delivered a withering reply to the latest German peace proposal, stating that the United States would never consent to an armistice "so long as the armed forces of Germany continue the illegal and inhuman practices which they still persist in" even as they were professing to seek peace. Finally, on October 20, the German government agreed to halt the U-boat campaign. Ludendorff would term "this concession to Wilson" the "heaviest blow" to the morale of the German armed forces.[28]

What actually was demoralizing the German armed forces was the increasingly obvious fact that Germany had lost the war. The government's delay in reaching a cease-fire agreement had in fact done exactly what Ludendorff had earlier warned it would, strengthening the Allies' hand as Germany's military and political situation crumbled. At the end of October, German battleship crews, ordered to steam forth in a final "death or glory" attack on the British Grand Fleet, mutinied. Soldiers sent to put down the rebellion joined the mutineers, and when the crews of the mutinying ships were broken up and transferred to other bases in an effort to stem the trouble the effect was only to spread the revolt throughout the fleet. The Kaiser was told by his ministers that the only hope now for averting a general revolution and saving the empire was his immediate abdication. "I have no intention of quitting the throne because of a few hundred Jews and a thousand workmen!" the Kaiser retorted. "Tell that to your masters in Berlin!"[29] He gave in on November 9 as the German delegation at Compiègne was preparing to accept the stern terms demanded by the Allies and a revolutionary mob in Berlin was proclaiming a socialist republic from the steps of the Reichstag. Ludendorff's successor at the high command explained to His Majesty that he no longer had an army that would obey his orders.

It was Haig, almost alone among the Allied military and political leaders, who had expressed grave reservations about the wisdom of extracting humiliating concessions from Germany. "I think this is a mistake," he wrote his wife, "and may encourage the wish for revenge in the future." Haig was especially dismissive of the Admiralty's reasoning that, since the German High Seas Fleet would surely have lost all of its modern warships had it ventured forth to do battle, it should therefore be required to turn over the entire fleet as one of the Armistice conditions.[30] His was a voice in the wilderness.

Within a year of the war's end Ludendorff and Hindenburg were insisting without batting an eye that a valiant and undefeated German army had been "stabbed in the back" by a weak and treasonous civilian government, abetted by socialists and Jews. "One had to live in Germany between the wars," wrote the newspaper correspondent William Shirer, "to realize how widespread was the acceptance of this incredible legend by the German people"—and how much it would drive Germany's resurgent militarism.[31]

3

Cambridge

PATRICK BLACKETT ARRIVED in Cambridge on January 25, 1919, and stepped into a new world. Promoted to full lieutenant the previous May, he was one of 400 junior officers the British navy decided to send to the university for a six-month course at the end of the war "with the object of instilling into us some general culture which had been lacking among those who had been whisked to sea in 1914 when very young," Blackett wrote. The officers were parceled out among the colleges of the ancient university and attended lectures in uniform, a striking enough picture that Rudyard Kipling was inspired to capture it in a slightly satirical (and largely forgotten) poem:

> Oh, show me how a rose can shut and be a bud again!
> Nay, watch my Lords of the Admiralty, for they have the work in train.
> They have taken the men that were careless lads at Dartmouth in 'Fourteen
> And entered them at the landward schools as though no war had been.
> They have piped the children off all the seas from the Falklands to the Bight,
> And quartered them on the Colleges to learn to read and write![1]

His very first night at Magdalene College, Blackett stayed up late talking with two other new students he had just met and who would become lifelong friends: Kingsley Martin, the future editor of the leftist *New Statesman*, and Geoffrey Webb, who would become one of England's foremost art historians. Martin's and Webb's discussion that night about God, Marx, and

Freud were, Blackett later recalled, the first intellectual conversation he had ever heard.[2] Martin's accounts of war in the trenches—a socialist and the son of a Congregationalist minister, Martin had declared himself a conscientious objector and been assigned to an ambulance unit in France—also made Blackett realize "how relatively comfortable the war at sea had been compared to the grim horrors of the Western Front."

Blackett had already had doubts about staying in the navy. During the last months of the war, assigned to a destroyer in Harwich Force, he had begun to read science textbooks with the idea of possibly getting a job with a scientific instrument company or going to university. He summarized his state of mind:

> I enjoyed my four years at sea during the war, but I was very doubtful if I would enjoy peace time Navy. There seemed to me to be two attitudes which I might take if I decided to stay in the Navy. I could treat the Navy as providing a pleasant way of life and an introduction to the best clubs around the globe, or I could take the technological problems of naval warfare very seriously and so become orientated towards fighting another war. As I put it to myself rather crudely: I enjoyed shooting at the enemy during the war—would I enjoy shooting at targets? I decided I wouldn't.

A few days after arriving at Cambridge, Blackett wandered over to the university's Cavendish Laboratory "to see what a scientific laboratory was like."[3] Three weeks later he marched into the office of his commanding officer at Cambridge, Commander H. E. Piggott, and announced that the "intellectual life of a place like Cambridge" was what he was cut out for; he couldn't see himself going back to the peacetime navy, spending the rest of his life "walking up and down with a telescope under his arm." Piggott replied that he did not see how he could further Blackett's plans, "seeing that this would mean depriving the service of one who would likely prove one of her brightest senior officers," but that if he was determined he should talk to his tutor at Magdalene and submit his application for resignation from the navy through the usual service channels.[4] Blackett did so at once, and never looked back. He promptly enrolled as a regular undergraduate studying for the intensive Mathematical Tripos examination; he passed Part I in May, then switched to physics the following fall. In 1921 he received a first in physics and was accepted as a research fellow at the Cavendish, whose new director was the towering figure of the physics world, Ernest Rutherford.

In a scant half century since its founding in 1874, with a bequest from the Duke of Devonshire, the Cavendish had become the preeminent physics laboratory in the world, catapulting British science to the forefront of the exciting new fields of radiation and atomic physics. Its first four directors were, and still are, legendary names in the history of modern science. James Clerk Maxwell, the first director, formulated the basic laws of electromagnetism that laid the basis for modern electronics and communications technology. Lord Rayleigh, his successor, made pioneering discoveries in light and sound and was the discoverer of the element argon. Rutherford's immediate predecessor, J. J. Thomson, had discovered the electron.

Rutherford arguably surpassed them all. Born to a homesteading family struggling to make a go of life on the rough frontier of New Zealand, Rutherford combined brilliance, ambition, and apparently inexhaustible energy to rise to this preeminent position in British science. He had won a series of scholarships as a young student, culminating in the prestigious 1851 Exhibition scholarship that brought him to Cambridge. He swiftly made a name for himself with original work on radio waves, switching fields abruptly after the discovery of radioactivity in 1896; the next year, at age twenty-seven, he became a professor at McGill University in Canada, where he proceeded to carry out groundbreaking studies of the transmutation of elements via radioactive decay which won him the Nobel Prize in chemistry in 1908.

Unlike virtually every other Nobel Prize winner in history, Rutherford then went on to make his greatest scientific discoveries *after* the work that won him the prize. Success never spoiled him; he was unstoppable. A colleague who had been a fellow student at the Cavendish with Rutherford was asked many years later if he and Rutherford had become friends back then. He replied, "One can hardly speak of being friendly with a force of nature."[5] Perhaps more to the point was what another colleague observed: Rutherford never lost "his genius to be astonished."

In 1911 he announced the results of some experiments that astonished everyone. The accepted model of the atom at the time pictured electrons evenly distributed through a cloud of positive charge. Rutherford had tried to probe the structure of this atomic space by shooting a stream of heavy alpha particles through a thin layer of gold foil and then measuring the small deflection of the particles from their course as they emerged on the other side. A thin layer of zinc sulfide spread on a glass plate served as a detector; alpha particles striking the coating created tiny, glowing tracks— "scintillations"—that could be seen through a microscope and individually counted. Looking for the tracks was tedious and difficult work. At one point, out of what his assistant thought an excess of experimental thoroughness, Rutherford suggested placing the zinc sulfide detector on the same side of the gold foil target as the alpha particle source. To his amazement, the detector glowed on that side, too: some of the alpha particles were rebounding directly off the target. "It was," Rutherford said, "quite the most incredible event that has ever happened to me in my entire life. It was almost as incredible as if you fired a 15-inch shell at a piece of tissue paper and it came back and hit you."[6] He realized that the only way to explain such a ricochet was if the entire positive charge of the atom were concentrated in an extremely small space, creating an enormous repulsion force as the positively charged alpha particle passed near it. He had discovered the nucleus—and the structure of the atom itself.

In 1919 Rutherford nearly equaled that monumental discovery in an experiment in which he bombarded nitrogen gas with a beam of alpha particles, chipping off a stream of protons from the target atoms. Rutherford was at the time serving on a committee of scientists exploring methods of detecting submarines underwater by sound—what would be the genesis of sonar—and he wrote Karl Compton, the chairman of the committee, a letter of apology explaining that he would be late for their next meeting, scheduled to take place in Paris: he had apparently just split the atom, and was in the midst of carrying out a second experiment to confirm the result. "If this is true," he wrote Compton, "it is a fact of far greater importance than the war."[7]

The Cavendish was remarkable not just for its scientific preeminence but for its existing at all in the ivory-tower world of Cambridge, with its traditional disdain for the practical, much less the mechanical. The laboratory occupied a nondescript, three-story, gray stone Victorian building in a crooked medieval alley that ran behind one of the older colleges. Its facade, in the words of one observer, could have "graced any Scottish hotel." The building bore no clue to its identity beyond a statue of its benefactor bearing the Latin inscription *Magna opera Domini exquisita in omnes voluntates ejus*: "The works of the Lord are great, searched out by all who have delight in them." Inside, the building was even less prepossessing, "uncarpeted board floor, dingy varnished pine doors and stained plastered walls, indifferently lit by a skylight with dirty glass."[8]

The physicist Max Born, who would flee Germany and the Nazis for Cambridge in 1933, noted with amusement the British academic gentility that insisted on calling theoretical physics "applied mathematics"; by analogous reasoning, he suggested, the Cavendish should be called the Department of Applied Glass Blowing.[9] The allusion to glassblowing was not a joke, though: every student and researcher at the lab was expected to master such practical hands-on skills, and it was part of the indoctrination of all new members of the lab to put in a stint in the "Nursery," a course for newcomers held in a cramped attic room filled with bits of miscellaneous equipment, where they learned the basic techniques. There was clearly a sort of in-group pride among the physicists in possessing skills so different from their fellow rarefied academics. Patrick Blackett, in an essay he would write upon his departure from Cambridge—titled "The Craft of Experimental Physics," it remains a classic exposition—underscored the satisfaction he and his fellow physicists took in being able to combine the mental and the manual in their daily work:

> The experimental physicist is a Jack-of-All-Trades, a versatile but amateur craftsman. He must blow glass and turn metal, though he could not earn a living as a glass-blower nor even be classed as a skilled mechanic; he must carpenter, photograph, wire electric circuits and be a master of gadgets of all kinds; he may find invaluable a training as an engineer and can profit always by utilising his gifts as a mathematician. . . . The experimental physicist must be enough of a theorist to know what experiments are worth doing and enough of a craftsman to be able to do them. He is only pre-eminent in being able to do both.[10]

Every year the research students at the Cavendish held a somewhat raucous celebratory dinner, the highlight of which was the performance of Gilbert and Sullivan songs and other familiar tunes with words reworked for the occasion. Published in a privately printed series of volumes (*Postprandial Proceedings of the Cavendish Society*), they were an affirmation of the comradeship of their world apart, often containing allusions to their experimental improvisations. To the tune of "Clementine," one year they sang:

> *In the dusty lab'ratory*
> *'Mid the coils and wax and twine,*
> *There the atoms in their glory*
> *Ionize and recombine.*[11]

"Wax and twine" was not just a metaphor, and the wax referred to was indeed something of a legend in its own right: the physicists had discovered that the red sealing wax used by the Bank of England was the perfect medium for sealing vacuum apparatuses, and there was scarcely an experiment in the whole lab that did not use it (until another improvisation, plasticine, displaced it).

The physicists were different in another way for being an unusually cosmopolitan group amid the generally still very insular society of British academia. The Cavendish regularly welcomed students and visiting scientists from around the world—Robert Oppenheimer from America, Peter Kapitza from Russia, Niels Bohr from Denmark among them—and everyone traveled and maintained a far-flung network of scientific contacts, in Germany, Denmark, Holland, Italy, France.

But even the physicists and the Cavendish were not completely immune to the genteel charms of an ancient university, and there remained something quaintly civilized about the rhythms of life at the world's preeminent experimental physics factory in the 1920s. For eight weeks a year, during university vacations, the doors of the Cavendish Laboratory were simply shut and locked; everyone was expected to go away and take a holiday, or catch up on some reading, or perhaps attend a scientific conference, but no work was done at the lab during these times, period. Every afternoon at 4 p.m. the researchers would drift down to a small room next to the library and gather for tea and cakes. (The tradition had been for the director's wife to supply these herself but Lady Rutherford finally drew the line as the staff grew, and the scientists were forced to set up their own "bun fund" to pay for the cakes.) Starting hours at the Cavendish were loose and unenforced, but at 6 p.m. sharp the workday ended, a rule sometimes emphasized by a technician going from room to room pulling out plugs and snapping off switches. A visiting American physicist was amazed at the leisurely pace of life there; he confessed in a letter to J. J. Thomson after his visit that "a laboratory in this country in which nobody ever began work before 10 a.m. or worked later than six in the evening" would be viewed as a "terrible example of sloth and indolence."[12]

It was a gentler era altogether and Cambridge was still an Eden with its own stately pace and splendid isolation from the rough-and-tumble world. Students still wore academic gowns to class, in the library, when calling on senior members of the university, and when walking anywhere outside the college after dark (where a mortarboard cap was required, too, and being caught without this proper dress in town by a university proctor was a punishable infraction). Sport was still distinctly amateur, and an hour or two of tennis, cricket, rugby, bicycling, or running part of the daily afternoon ritual. Dinner every evening in the college halls was a formal and unhurried affair invariably including soup, fish, entree, roast, sweet, and savory; politics was rarely discussed.[13]

"TO BE DEPOSITED, so to speak, on the shores of Cambridge just as the Cavendish Laboratory was rising under Rutherford's inspired direction to great heights of eminence was luck indeed for me," Patrick Blackett would later write. "So I owe much to the Royal Navy."[14] He had learned of his research fellowship shortly after his graduation in 1921 when he received a letter from the president of Magdalene inviting him to a dinner at college. There was a railroad strike on and he was at his parents' home in Surrey but he got on his bicycle and pedaled to Cambridge, where he was told that both he and Kingsley Martin had been elected to fellowships by the college. "The world was at our feet," Martin remembered thinking at the news, "an oyster to be opened with the sharp sword of Cambridge intellectualism."[15]

Blackett had the advantage, too, in this new world he had been deposited in, of making a striking impression, both physically and personally. Ivor Richards, a Cambridge don and literary critic who would become one of the most influential founders of the New Criticism movement, was Geoffrey Webb's adviser and through him met his friend Blackett; years later Richards would vividly recall the day Blackett first stopped by his flat, which was at the top of a dilapidated building at the end of the same lane where the Cavendish stood. "Came a quick step . . . a tap on the door, and there entered a young Oedipus. Tall, slim, beautifully balanced and looking always better dressed than anyone. . . . But above was that mysterious intense and haunted visage. . . . The tragic mask, however, was highly mobile, alive indeed with intelligence, modesty and friendliness." Solly Zuckerman, a zoologist who met Blackett a few years later, had a very similar reaction: "tall and strikingly handsome in the film-star mould," yet "measured," "not immediately forthcoming. . . . He often looked as though all the cares of the world were on his shoulders."[16] Late in Blackett's life a colleague remarked to him that his official portrait as president of the Royal Society made him look very serious. "But I am a very serious man!" he replied.[17]

Every new arrival at the Cavendish was conscripted to the tedious task of counting scintillations in Rutherford's ongoing bombardment experiments. Blackett remembered hours spent in a dark room, letting his eyes adjust, and then straining to look for the faint tracks for a minute at a time as the experiment proceeded.[18] But Rutherford also firmly believed in helping his students and research fellows learn to carry out their own research projects: it was why so many of the world's top-rank physicists had gotten their start at the Cavendish under Rutherford. Even as he kept up his own experiments and growing administrative tasks, the director made the rounds of the laboratory every morning at 11 a.m., singing "Onward Christian Soldiers!" off-key as he came down the corridor, his voice booming as he asked questions, offered suggestions, and dispensed encouragement, or with alarming frequency erupted in unpredictable but mostly harmless rages that "his boys" learned to shrug off. "We are living in the heroic age of physics!" he declared at the annual meeting of the British Association for the Advancement of Science, and few of his protégés doubted it.[19]

Within a year Blackett had published his first scientific paper based on his own research and was well on his way to the discoveries that would catapult him to the top ranks of physics, with a speed remarkable even by Cavendish standards. Blackett would later say that he never forgot this early lesson Rutherford gave him, of the importance of "choosing the really important problems and letting a young man get on with them."[20] Rutherford had suggested that Blackett try to adapt the cloud chamber—a device that, with characteristic enthusiasm, Rutherford had declared to be "the most original and wonderful instrument in scientific history"—to more precisely document the artificial disintegration of the nitrogen atom when it collided with an energetic alpha particle.[21]

The cloud chamber had been invented twenty-five years earlier by another Cavendish scientist, C. T. R. Wilson. His lectures on light, which Blackett had attended, were legendary at the laboratory equally for their brilliant treatment of the subject and for Wilson's comically abysmal presentation, delivered in barely audible tones with his back to the audience, his faint writing on the blackboard almost immediately wiped out by the eraser he held in his left hand as he worked his way across the board. In 1894 Wilson had spent a few weeks atop Ben Nevis, the highest mountain in Scotland, where a small observatory had been established, and he had become intrigued by "the wonderful optical phenomena" known as coronas and glories, produced when the sun shone through the clouds surrounding the hilltop. Hoping to reproduce these effects in the laboratory, he built an apparatus that could be filled with a mixture of air and water vapor. A piston allowed the volume to be suddenly expanded, lowering the temperature and causing the water vapor to become supersaturated; at that point the vapor would begin to condense into droplets, forming a visible cloud. But as Wilson said, "Almost immediately I came across something which promised to be of more interest than the optical phenomena which I had intended to study." The condensation of supersaturated water vapor into visible droplets begins at so-called condensation nuclei, typically specks of dust. But when Wilson carefully removed dust from the air in the chamber, he noticed entire trails of droplets forming in the supersaturated vapor mixture. His great discovery was that it was charged particles that were causing the phenomenon; when he placed a radioactive source next to the chamber, a cloud trail tracing the path of the charged radioactive particles vividly appeared. He subsequently was able to produce beautiful photographs of the tracks of alpha, beta, and X-ray particles. Wilson would win the Nobel Prize for physics in 1927 for this work.[22]

Rutherford at first gave the job of adapting Wilson's apparatus to a Japanese student, who subsequently had to return home suddenly for family reasons. The project was passed on to Blackett. "So I found myself with a few bits of Schimitzu's apparatus in an otherwise empty research room and told to get on with it," Blackett recalled.[23]

Blackett would become legendary among his fellow scientists for his ability to combine physical insight, mathematical understanding, and extraordinary practical skills, and from this very first independent project of his own he displayed what one colleague would call his "remarkable facility" of "thinking most deeply when he was working with his hands." Another colleague said: "He could use physical theory to design a piece of equipment and then draw it, make it and get it to work. I have known nobody who possessed this combination of talents in the degree that he did."[24]

The trick of capturing good images with the cloud chamber was to create a sharp, rapid expansion and then instantly take a photograph before the tracks faded. Blackett ingeniously solved the problems by designing an automatic device that mechanically linked the piston and the camera; to keep the chamber clear of old fading tracks, the radioactive source (a polonium deposit on the end of a copper wire) was shielded by a mechanical shutter also linked to the expansion bar, so that the alpha particles only entered the chamber at the moment sufficient supersaturation occurred, ensuring the tracks would be sharp.

With this automated mechanism Blackett was able to produce a thousand photographs a day. He eventually had a staggering 23,000 images, each containing an average of eighteen alpha-particle tracks. A small number of the tracks made a sudden fork—a clear indication of a collision with a nitrogen or oxygen atom. But a grand total of eight out of the half a million tracks he had captured and pored over were the Holy Grail. These, as Blackett reported, were "of a strikingly different type. These eight tracks undoubtedly represent the ejection of a proton from the nitrogen nucleus." The surprise was that instead of three tracks emerging—the ejected proton, the residual nucleus, and the alpha particle—there were only two; the alpha particle was apparently being captured in the reaction. The angle and momentum of the tracks allowed Blackett to calculate the mass of the new atom that had been produced in the process, and they left no doubt that it was an oxygen atom: "In ejecting a proton from a nitrogen nucleus the alpha-particle is therefore itself bound to the nitrogen nucleus."[25] Blackett had captured the first photograph of a nuclear reaction as it actually took place.

In little over two years he had not only fulfilled the task Rutherford had assigned him but achieved a triumph at the forefront of physics. His cloud chamber image of a nitrogen atom being transformed by a nuclear collision into an oxygen atom would become famous—literally—when it was published the following year, 1925, in a paper in the *Proceedings of the Royal Society* reporting the work. A tribute to Blackett that appeared on his death in 1974 noted that this photograph had appeared in virtually every physics textbook in the half century since. "Patrick's transformation from a young naval officer who had been under fire into a university man whose scientific genius immediately became apparent was, I should say, unique," said Solly Zuckerman. "I certainly do not know of any similar case in the academic history of Great Britain."[26] It was a measure of his precocious success that he never bothered to obtain a Ph.D. degree, a fact that he took slightly wry pride in throughout his life.[27]

IN MAY 1924 Blackett married Costanza Bayon, a free-spirited and unconventional young woman who was studying modern languages at Cambridge. The daughter of an English mother and an Italian father, she had been raised by an English couple in Florence who had renamed her Dora Higgs. On arriving in England she rebelliously resumed her old name. But she was known to everyone by the nickname she had given herself in childhood, Pat. To their close friends, the newlyweds were "the two Pats."[28]

They spent their honeymoon in Italy that spring; eager to return to the continent for a longer sojourn, Blackett proposed to Rutherford that he take a sabbatical in the 1925–1926 academic year at the University of Göttingen in Germany. Göttingen had emerged as a leading center of the new quantum theory, and the scientists who would pass through its Institute for Theoretical Physics as students, visiting lecturers, or faculty in the 1920s would be a veritable Who's Who of twentieth-century nuclear science: Robert Oppenheimer, Niels Bohr, Werner Heisenberg, Enrico Fermi, Edward Teller, John von Neumann, Eugene Wigner, Wolfgang Pauli. Heisenberg and Max Born, the institute's director, had just developed the matrix mathematics representation of quantum mechanics that would win them each a Nobel Prize in physics (though twenty-two years apart). But quantum mechanics focused more on the electronic structure of the atom than on its nuclear core, and Rutherford was not enthusiastic about the idea of Blackett's proposed sabbatical. "I remember vividly the rather grudging permission from Rutherford for me to leave the Cavendish for a year (my first sin) and to study the outside of the atom rather than the nucleus (my second sin)," Blackett wrote.[29] It was the first indication of a perhaps inevitable rift between the two men. Blackett spent the year at Göttingen working with the experimental physicist James Franck, who would receive the Nobel Prize that year for his work on the electron, and who would later be one of the first prominent academics in Germany to protest the Nazis' racial laws by resigning his position in April 1933. A colleague of Blackett's at the time of the Nazi takeover would recall him speaking of his year in Germany and "his strong affection for its people and its culture . . . listening to him one could feel all his nostalgia for a Germany he could no longer accept."[30]

Blackett had never given much thought to politics. His father had been a Liberal who drifted Conservative in his later years; Patrick had voted Conservative in the 1918 election, he later said, because it was the "natural" thing for a naval officer to do.[31] His service at sea during the war had already begun to plant doubts, though. He would later tell a colleague that he had instinctively disliked the class distinctions maintained between officers and men, and the rituals intended to reinforce them; he had found particularly ridiculous the way he had been taught as an officer to walk several yards away from the men before turning and shouting his orders to them.[32] The war, and even more the first few years of peace, were a radicalizing experience for many Britons as their hopes for a better world were inevitably disappointed. The historian Robert Wohl in his book *The Generation of 1914* notes that "the famed cynicism and disillusionment of the survivors" of the trenches "were, to a great extent, a product of the first few years of peace," and of the unrealistic expectations soldiers had brought back with them:

> Many soldiers had . . . come to think that the war must have a secret mean-
> ing that only the future would reveal; they found it necessary to believe
> that their sacrifice and suffering would not be in vain; and they clung to
> the hope that the war would turn out to have been a rite of purification
> with positive results. . . . It was said, and widely believed, that class barriers
> would fall; that selfishness would give way to cooperation; that harmony
> would reign; that conflict among nations would cease; and that everyone's
> sacrifice and suffering would somehow be compensated.[33]

The year 1919, the army veteran and war poet Siegfried Sassoon observed, "laboured under a pervasive disadvantage. Too much was expected of it. It was a year of rootless rebeginnings and steadily developing disillusionments."[34] Great Britain entered the war as one of the world's major suppliers of coal, ships, textiles, and iron and steel. The cotton mills of Lancashire produced half the world's yarn and cloth, the shipyards of the northeast a third of the world's ships. These traditional industries accounted for three quarters of Britain's exports and a quarter of all jobs. The war not only cost Britain £11 billion and transformed her from a creditor to a debtor nation but saw the permanent loss of many important markets to producers in the United States and Japan. Industries such as shipbuilding that had boomed to meet the demands of war crashed in the first few years of the peace. A speculative craze in the textile industry led to 42 percent of the Lancashire cotton spinning capacity changing hands in 1919 and 1920, followed by a similar implosion. In 1922 exports of cotton textiles were less than half their immediate prewar levels. Coal exports were one third the previous figure. Unemployment hit 2 million by the summer of 1921. The Wall Street Crash of 1929 and the Great Depression that followed were, in Britain, less a bolt from the blue that brought a halt to a boom of the Roaring Twenties than a continuation of "the Slump" that had afflicted the country ever since the end of the war.[35] As the writer and war veteran Robert Graves put it, for returning soldiers the standard topic of conversation simply changed from "this bloody fucking war" to "this rag-time fucking peace."[36]

For much of Britain's working class, the world they returned to remained gray, closed, and grim. The government promised "homes fit for heroes," but millions of workers in Glasgow, Liverpool, Manchester, and other urban slums continued to live in Dickensian squalor, crammed two or more to a room in run-down tenements with a shared water tap and WC in the backyard. (A Liverpool tenant who moved into new government housing in the 1930s marveled at realizing the "dream" of indoor plumbing: "Having a toilet next to your bedroom, especially in winter, you felt like the Queen.") Even in areas that had escaped the worst of the slump, seasonal unemployment was a fact of life as builders, seaside resorts, and even carmakers and other manufacturers regularly laid off thousands of workers every winter as demand slackened. Two and a half million men, permanently disabled by the war, received meanly computed pensions based on the precise extent of their injuries (loss of an entire right arm was worth 16 shillings a week, but only 14s. if the arm had been severed between the elbow and shoulder, 11s. 6d. if below the elbow), relegating many once skilled workers to begging or jobs barely much better, selling newspapers, bootlaces, and matches on the street. In more hard-hit regions, particularly the coalfields and industrial centers of Scotland, Wales, and northern England, entire communities packed up and migrated to London, the Midlands, and the South seeking work. Working-class families trying to make ends meet crowded pawnshops; a Lancaster factory worker remembered "queues a mile long on a Monday morning" as women waited to pledge items for a few odd shillings.[37]

Working-class fertility rates declined dramatically as married women sought to limit the size of their families, but ignorance and embarrassment about birth control methods remained widespread. The most commonly used method by married couples was still withdrawal, which surely added to the frustrations of returning soldiers for whom the war in France had offered at least a peek at more adventurous sexuality. (A young newly married man recalled receiving coyly phrased advice about coitus interruptus as contraception from a bus conductor friend, who admonished him, "Don't forget, always get off the bus at South Shore, don't go all the way to Blackpool.")[38]

But throughout Britain, not just among working stiffs and their gloomy privations, there was a powerful sense of dislocation, of society having come adrift from its familiar moorings: a sense that something needed radical fixing. "The old order is doomed," declared the Duke of Marlborough. He was commenting specifically about the millions of acres of venerable estates broken up and sold in the first few years after the war, in part a result of the higher taxes and death duties to pay for the war. He might have been talking about everything. The poet Richard Aldington recorded the disconcerting sensation that the war had left Britain a meaner place:

> In a very short time I realised that the London I had come back to was a very different place from the London I had left in 1914, let alone prewar London. Everything seemed askew. The streets were dirty and shabby—there were no men to clean them and nothing had been repaired or repainted for years. There were holes even in the main thoroughfares. The decent, orderly, good-natured Londoners had become as snappy and selfish as the far more sorely tried French. There was a shortage of everything except returning soldiers and debts. People fought for places in the inadequate transport system—a man who was accustomed to make way for women could not get on a bus. Food was scanty and very dear. Lodgings or apartments were almost impossible to find, because London was crowded with enormous numbers of "war workers," who still clung to their jobs like limpets. There was a devil-take-the-hindmost scramble for money and position in the new world, and an extravagance which seemed incredible to me who had known the old sober England. I stood aghast at this degeneration of my people.[39]

A small straw in the wind was the first appearance, in 1919, of P. G. Wodehouse's Bertie and Jeeves; Wodehouse's "lyrical-ludicrous style" (Robert Graves's description) and the extraordinary gentleness of his uproarious humor could not conceal the fact that this satire upon British upper-crust society would have fallen flat a generation earlier.[40] Much space was devoted by that reliable voice of the establishment *The Times* to pointing out the many ways, large and small, that everything was going to hell in Olde England. A particular case in point was the invasion of "the jazz," which *The Times* called "one of those American peculiarities which threaten to make life a nightmare," and which a certain Canon Drummond of the Maidenhead Preventive and Rescue Association denounced as "one of the most degrading symptoms of the present day . . . the dance of low niggers in America."[41] The first London theater season since the return of peace had created the expectation of "better plays," *The Times* predicted, now that theatrical managers could no longer count on the wartime crowds of London visitors who had reliably filled their theaters no matter what was on. But the autumn of 1919 brought a wallowing in nostalgia, revivals, and old standards that itself said something about a public looking hard for reassurance: there was Gilbert and Sullivan (*Iolanthe* and *The Gondoliers*) at the Princes Theatre, *Cyrano* at the Duke of York's, and Shakespeare at the Aldwych, the Old Vic, and the Court—the last presenting a *Merchant of Venice* whose Shylock was portrayed by one Mr. Moscovitch ("said to be a Russian Jew") with such fidelity to every hoary anti-Semitic stereotype as to earn *The Times* reviewer's effusive praise along with the observation, "We do not excuse, but begin to understand, pogroms."[42]

The tragic myth of a "lost" or "missing" generation that had been created by the war, and the concomitant belief that the war itself had been a sort of cruel and cynical joke perpetrated on the patriotism and courage of a doomed and idealistic youth, would actually arise only about ten years later in a spate of memoirs, novels, and plays in Britain. It became a way of explaining the disappointments, romanticizing the past, assuaging the guilt of the survivors.[43] The war dead amounted to 7 percent of the male population aged fifteen to forty-nine in England and Wales: a terrible toll, but hardly the elimination of an entire generation, or even "the best" of a generation.

There was nonetheless a kernel of truth to the feeling that the war had taken a disproportionate toll on the educated elite. "The emphasis on the recruitment of 'gentlemen' into the officer corps at the outbreak of the war . . . tended to draw the officers from the existing social elites," notes the social historian James Stevenson. The good health of members of the upper classes compared to that of many manual laborers, moreover, made it far more likely that they ended up in frontline infantry units: "Of the 13,403 students from Oxford who served in the war, 2,569, almost one in five, were killed. Cambridge showed virtually identical figures (13,126 served; 2,364 killed)."[44] At the universities, a sense of obligation to make that sacrifice mean *something* intruded insistently on the prewar tranquillity of splendid scholarly isolation.

CAMBRIDGE IN THE 1920S was far from a hotbed of left-wing politics. Rutherford pointedly frowned on political activity, believing that science, and scientists, should stay out of the political arena (and public controversy) altogether.

But through those friends of his first night at Magdalene, Kingsley Martin and Geoffrey Webb, Blackett was introduced to a growing circle of prominent left-wing intellectuals. Notable among them were Bertrand Russell, George Bernard Shaw, the Marxist political economist Harold Laski, the psychiatrists Adrian and Karin Stephen (Virginia Woolf's brother and sister-in-law) and W. H. R. Rivers (a Cambridge don who had pioneered the psychiatric treatment of "shell shock" and figured prominently in the 1917 saga of Siegfried Sassoon when that by then much decorated war hero refused to return to the front as a protest), and the zoologist Solly Zuckerman. Through Ivor Richards, Blackett and Martin became enthusiastic members of a Cambridge society known as the "Heretics," who met for discussions and debates about philosophy and art and who were dedicated to the proposition of rejecting "all appeal to Authority in the discussion of religious questions." The Heretics also brought Blackett into the Bloomsbury orbit, including Virginia and Leonard Woolf, Roger Fry, and Lytton Strachey.

The 1926 General Strike would prove a watershed in the emerging political consciousness of the British intellectual left in the 1920s. The strike was the culmination of a long developing crisis that brought together many of the festering ills of Britain's postwar social and economic malaise. As one of his first acts as the new chancellor of the exchequer in 1925, Winston Churchill, without any public debate, announced that he had returned Britain to the gold standard. The immediate effect was to increase the cost of British exports by about 10 percent, a disaster for the coal mining industry in particular. The mine owners declared they would have to cut wages; the 1,250,000 miners threatened to go on strike, and the Labour Party and the umbrella Trades Union Congress pledged to support them; at the last minute the government announced that it would subsidize the mine owners to maintain wages while a commission studied the problem.

That only postponed the inevitable confrontation. The commission reported the following spring that the owners had taken huge profits and failed to invest in new equipment to keep the British mines competitive, but concluded that in the short run the only alternatives were a wage cut, or indefinite continuation of government support. To Prime Minister Stanley Baldwin it was self-evident that it was the workers who needed to make the sacrifice for the greater good: "All the workers of this country have got to take reductions in wages in order to help put industry on its feet," he said. On May Day, 1926—impropitious timing if ever there was—the mine owners announced that unless the workers agreed to an immediate reduction in pay they would be locked out. Two days later the Trades Union Congress declared a general strike of all 6 million of its members, from printers to steelworkers to train porters.

The government responded with a call for volunteers, and thousands of middle- and upper-class men and women lined up to drive delivery trucks, operate power plants and gas works, and keep food supplies flowing. Buses crawled through the streets of London with barbed wire wrapped over their radiators and a policeman seated next to the driver, while armored cars and squads of soldiers guarded bus garages. In Hyde Park dozens of huts served as the headquarters of a huge operation to receive and distribute milk.

Many in the Cambridge and Oxford communities volunteered to serve with the strikebreakers (there is a minor but memorable scene in Evelyn Waugh's novel *Brideshead Revisited* of Charles Ryder and his Oxford friends forming a "high-spirited, male party" that signs up for duty with a volunteer militia, convoying milk deliveries around London and "looking for trouble"). Blackett was one of the few members of his university who visibly took a stand on the other side, driving to London to pick up and distribute copies of the strikers' newspaper, the *British Worker*.

After ten days the strike fizzled out. Churchill, who had built a remarkably progressive record on social policy in his years in public office even as a member of Conservative governments, had been strongly sympathetic with the miners and supported their right to strike; he had confidentially dispatched fellow Conservative MP Harold Macmillan to Newcastle to investigate the situation and had been deeply moved by Macmillan's account of the deplorable living conditions and suffering of the workers and their families. But he saw a general strike as a revolutionary challenge to the rule of law and was determined to see it crushed. With the newspapers not publishing, the government stepped in to print its own official paper; if the *British Gazette* was little better than propaganda it was largely because of the man who leapt in to edit it. Churchill threw himself into the job with his trademark gusto. "He butts in at the busiest hours and insists on changing commas and full stops until the staff is furious," noted one bemused observer. But he also dictated lengthy editorials and ordered up stories accusing the strikers of fomenting revolution, referring to them as "the enemy," calling a Labour MP who spoke up to support the strike "a wild Socialist." Kingsley Martin triumphantly declared that Churchill had been "discredited" once and for all by this fusillade of belligerent and often dishonest rhetoric. The episode left a residue of bad blood between Churchill and the intellectual left that never completely vanished.[45]

If Blackett was an early convert to socialism, he was nonetheless part of what would become an unmistakable trend in Britain's scientific community between the wars. Benjamin Farrington, an Irish classicist and Communist Party member who returned to London in the mid-1930s after teaching in South Africa for several years, said he had the impression that "at least half the Marxists whom I met were scientists." The writer and scientist C. P. Snow estimated that three quarters of the 200 "brightest" young physicists in Britain in the interwar years would come to view themselves as on the political left.[46]

The reasons were complex. Most scientists, to be sure, were instinctively antiauthoritarian to begin with; science was in principle an open republic in which all could contribute equally based only on the quality of their work, regardless of standing or status, and in which truth was arrived at as the natural outgrowth of a vast mutual enterprise rather than dictated by received authority. Science tended to attract men and women who by their natures had little patience with conventional wisdom, social elitism, or the veneration of authority and who embraced progress as part of their basic credo.

But it was the growing contradiction between the revolutionary excitement of scientific advance and the stagnation of society and the economy that radicalized many young physicists in this period. "Living in Cambridge," Snow would recall, "one could not help picking up the human as well as the intellectual excitement in the air." The discoveries pouring out of the Cavendish Laboratory were "part of the deepest revolution in human affairs since the discovery of agriculture," and the discoverers themselves were keenly aware of the fact.[47] Blackett would put his finger on exactly this point in the opening words of a broadcast he presented in 1934 on the BBC, *The Frustration of Science:*

> I think everyone will agree that the most striking fact about the present-day world is the contrast between the vast possibilities of prosperity and the appalling poverty of the majority of the population. Industry and science have made such huge advances that a large improvement in the standard of life, particularly of the workers, is now technically and immediately possible. But the social and economic structure of our western world is clearly of such a kind that we are unable at present to take full advantage of the technical progress which we have already achieved.[48]

The rest of the broadcast was a strange blend of sharp, original thinking and woolly cliches of Marxist class analysis and economic determinism ("it is very interesting to notice that the main popular support in any country for

Fascism lies in the lower middle classes and peasantry, and that these classes have always been effectively anti-scientific"). But his basic conclusion was that the full potential of science was inevitably stifled by the capitalist system and its pursuit of profit; even "planned" capitalism would never fully put science to work, certainly not for the greatest common good. Capitalism, by the same token, would never give science the full support it needed to make further progress:

> I believe that there are only two ways to go, and the way we now seem to be starting leads to Fascism; with it comes restriction of output, a lowering of the standard of life of the working classes, and a renunciation of scientific progress. I believe that the only other way is complete Socialism. Socialism will want all the science it can get to produce the greatest possible wealth. Scientists have not perhaps very long to make up their minds on which side they stand.[49]

On their return to Cambridge from Germany the two Pats had taken a house at 59 Bateman Street, one of a row of brick town houses facing the University Botanic Garden, about a mile from the Cavendish Laboratory. The couple were, thought Ivor Richards, "the handsomest, gayest, happiest pair in Cambridge." Once a week the Blacketts held an open house for their "semi-bohemian and left-wing" friends and colleagues, regaling them with lemonade, biscuits, and movies extolling the achievements of the Soviet state in collectivizing agriculture and advancing industrial production.[50]

"BOTH THE POLICY OF APPEASEMENT supported by the Left in England and the policy of aggressive expansion adopted by Fascist Germany and Italy were expressions, in their own ways, of the . . . lessons learned on the battlefields of the Great War," observes Robert Wohl.[51] The growing propensity in Britain to view its victory in the war as a tragedy was matched by Germany's willingness to regard its defeat as a crime.

Karl Dönitz returned to Germany in July 1919 as the country was being roiled by a series of right-wing putsch attempts led by irregular *Freikorps* of army and navy volunteers, many secretly supplied and organized by old-line officers who remained loyal to the monarchy even after they had pledged their allegiance to the new republic. The young U-boat captain had spent the last month of the war and the months since as a British prisoner of war.

His last command, *UB-68*, was on patrol in the Mediterranean about 150 miles east of Malta when he had surfaced in the middle of a convoy to find destroyers bearing directly down on him. In the chaos of executing a crash dive to escape, the boat had gyrated out of control—either the ballast or the hydroplanes were mishandled—and the boat hurtled up and down several times, reaching a depth of 300 feet at one point and an angle of 45 degrees before rising again and breaking crazily above the surface stern-first. Dönitz gave the order to abandon ship and the crew leapt into the water just seconds before the boat went down for the last time. The engineer, who had gone below to open the seacocks to scuttle the boat, went down with her. Dönitz fell into a deep depression, blaming himself for the loss of the boat and the death of his engineer. He would later hint that he had feigned insanity in the prison camp to obtain an early release, but it may have been more than just good acting. In either case he was among the first German prisoners of war to be repatriated following the signing of the Versailles treaty.

The Social Democrats polled at the top in the first vote held for the new Reichstag but still fell far short of a majority; one of the chronic weaknesses of the new Weimar Republic was a complex system of proportional voting that ensured representation of small parties but undermined the emergence of a stable majority. In many ways, though, the real power of the state had already passed to the army. The high command had come weakly to the defense of the socialist government in crushing a communist uprising that attempted to declare a "soviet" in January 1919. In return the armed forces exacted what was in effect a free hand in military affairs—and, as would become clear only too late, political affairs as well. Their first order of business was evading the military terms of the Versailles treaty and beginning without delay to rebuild Germany's army and navy.

This was an agenda that enjoyed overwhelming popular support, even from the broad center of German public opinion and the nation's democratic centrist politicians. The progressive Jewish industrialist and statesman Walther Rathenau would be assassinated by right-wing hit men in 1922 while serving as foreign minister; yet he, a member of the social liberal German Democratic Party, had singlehandedly done more than probably anyone in Germany to lay the groundwork for the nation's secret rearmament. Two months before his assassination, he had personally drafted a secret protocol to the Russo-German peace treaty allowing Germany to clandestinely construct and operate a military air base in Russia. (Under the Versailles treaty, Germany was permitted no air force at all.) At Lipetsk, about 250 miles southeast of Moscow, the German government poured out millions of reichsmarks a year to build and operate a modern airfield complete with two runways, hangars, and machine shops. Over the next ten years the Germans secretly trained hundreds of fighter and bomber pilots, refined aerial tactics, tested new weapons, and practiced dropping live bombs, and even poison gas, on simulated targets.[52]

Germany's undimmed spirit of defiant militarism was hardly a secret, though; it was frequently on display, and often in deeply disturbing fashion, to anyone who cared to look. In 1921 one of the handful of war crimes trials that would take place in Germany came before the Supreme Court in Leipzig. The case involved a horrifying atrocity committed by the captain of *U-86*, Oberleutnant zur See Helmut Patzig, on June 27, 1918. Patzig himself was not in court; he had vanished. But his two watch officers were standing trial for their part in the incident. There was little dispute about the facts. The Canadian hospital ship *Llandovery Castle* had been torpedoed by *U-86* off the coast of Ireland. When Patzig learned the identity of the ship he had just sunk—he actually surfaced, approached several lifeboats, and interrogated the survivors—he attempted to sink the boats by running them down, then ordered his officers to open fire with the deck guns. After thus attempting to erase the evidence of his act, he had sworn the officers to secrecy and ordered the logbooks altered to place *U-86* at a point far from where the sinking had occurred.

The court concluded that the excellent military skills possessed by German officers made it all but certain that the victims had not survived: "The universally known efficiency of our U-boat crews renders it very improbable that the firing on the boats, which by their very proximity would form an excellent target, was without effect." Nor was it an acceptable defense that the officers were merely obeying orders; it was impossible for them not to have known that an order to fire on lifeboats was unlawful. Still, Patzig had clearly acted in a "state of excitement" and so "the deed cannot be called deliberate." The court found the two officers guilty of homicide and sentenced them to four years' imprisonment. Even that mild punishment provoked a public outcry and an outpouring of support for the men. Within six months both men "escaped" confinement, and that was the end of the business as far as Germany was concerned. As Dönitz's biographer Peter Padfield would observe:

There could not have been a better example of the mood in leading circles, nor of how the ground was already prepared for Hitler: patriotism, expressed as defiance of the former enemy powers, was a higher value than justice; mass murderers of medical staff including nurses served terms which would have been lenient for petty larceny, while the officer who gave them their orders went free.[53]

On his return to Germany, Dönitz was unsure whether he would stay in the navy, serving the new socialist republic, but several influential and well-placed connections let him know in so many words that they did not expect the current state of affairs to last long. His father-in-law, a general from a venerable Prussian family, told him he had decided to stay in the army, adding, "You are not permitted to abandon the State!"

More compelling was the conversation he had with his former U-boat flotilla chief, Korvettenkapitän Otto Schultze, upon reporting back for duty.

"Are you going to stay with us, Dönitz?" Schultze asked.

"Do you think we shall have U-boats again?"

"Certainly I think so," Schultze replied. "The ban will not remain forever. In about two years it is to be hoped we will have U-boats again."[54]

Dönitz stayed. Within two years he was a Kapitänleutnant serving in what was for all intents and purposes the shadow U-boat force of an already rebuilding German navy. Under the Versailles treaty Germany was permitted to have torpedo boats, and the navy exploited that loophole to the hilt to train and develop tactics that would be just as applicable to submarine warfare. In the winter of 1921–1922 a staff exercise studied the use of night surface torpedo attacks by U-boats as a promising tactic to counter the defensive advantage of convoys. ("The coming war may or may not involve war against merchant shipping," the study noted, but since an enemy warship was itself part of a defensive convoy, the same tactics would be applicable to that situation as well.) The torpedo boats also began practicing the technique of locating an enemy convoy by day, remaining at a distance until darkness allowed them to stealthily approach, then attacking and quickly escaping at high speed.

In April of 1922 a Dutch shipbuilding firm by the name of NV Ingenieurskantoor voor Scheepsbouw, or IvS, was established at The Hague. Its name translates as "Engineer Office for Shipbuilding." In fact it was a dummy corporation set up by three German shipyards to begin building submarines to designs developed in Germany. Other German U-boat designs would be built in Finland, Spain, and Japan over the next several years, ostensibly for these or other foreign navies but with intimate German technical involvement. The naval high command in 1926 selected for development several U-boat designs to meet the requirements for "Case A," a war with France and Poland. In the summer of 1930 a group of German "tourists" arrived in Finland for an extended holiday. They were the first contingent of active duty German naval officers to begin training in U-boats, and for the next three months they carried out trials on a new 500-ton Finnish submarine that had been designed by IvS, and completed in Finland under the supervision of German engineers.[55]

IN 1932 Blackett began the work that would lead to his winning the Nobel Prize in physics sixteen years later. It was work revealing both about his experimental and analytical intuition, and of a small but significant rigidity in his approach. Researchers in France and the United States had recently reported discovering cloud chamber tracks apparently created by cosmic rays randomly passing through the chamber at the moment the vapor mixture had been expanded and photographed. Cosmic rays were charged particles created in outer space, and by measuring the curvature of the tracks when a strong magnetic field was applied it was possible to calculate their energy and mass; most were apparently single protons or electrons.

Randomly snapping and developing cloud chamber photographs in the hopes that every now and then one might happen to coincide with the moment a cosmic ray passed through was obviously an inefficient way to conduct science. Giuseppe Occhialini, a young Italian physicist, had been doing some work on detecting cosmic rays with Geiger counters, and he arrived at the Cavendish for what was intended to be a three-week visit. He stayed for three years, collaborating with Blackett on a series of experiments that would in no small measure contribute to C. P. Snow's exultant declaration that "1932 was the most spectacular year in the history of science."[56]

Occhialini (known as "Beppe") instantly fell under the spell of Blackett and his combination of "the superlative artisan and the dedicated scientist," as Occhialini would describe him. He recalled Blackett as a scientific and a personal inspiration,

> working with great efficiency and joyful intensity in the lab all day: going
> home in the evening to study, or to sit by the fire, fondling the ears of his

sheepdog, Bo'sun, puffing his pipe and smiling patiently at the very confused English of a young Italian; striding out on Sunday mornings, rain or
no rain, over the Cambridgeshire grass, with Bo'sun and I racing hard to
keep up with his long legs. . . . For those of us who in one way or another
were facing the problem of fighting fascism at home, Blackett represented
a hope for the future, an assurance that we were not alone. . . . For me, and
for the many exiles he helped and befriended, Blackett was England, an
England that maybe only existed in his mind and ours, but which gave us
courage and hope.[57]

Their idea for improving the process of capturing cosmic ray tracks was a
melding of the techniques the two men had been working on independently.
"The feeling between two partners in research is close to that between two
mountain climbers on the same rope," Occhialini said of their collaboration. Two Geiger counters, one above and one below the cloud chamber,
were electronically linked to send an electric signal only when both detected
a cosmic ray, indicating that the ray was passing through the plane of the
chamber. The signal then operated a motor to expand the chamber and
operate the camera shutter. If everything worked right, a cosmic ray track
would show up on every photograph. "I can still see him, that Saturday
morning when we first ran the chamber," Occhialini wrote forty years later
about Blackett's reaction to that first test, "bursting out of the dark room
with four dripping photographic plates held high, and shouting for all the
Cavendish to hear, 'One on each, Beppe, one on each!' "[58]

One of Blackett's famous dictums to his students was, "You should treat
your research like a military campaign. . . . Make sure you gather plenty
of data!" With their automated system in place, Blackett and Occhialini
proceeded to do just that, taking more than 700 photographs over the following months. In a nomination letter to the Nobel committee in 1948, his
Cambridge colleague J. D. Bernal would observe of Blackett, "Two features
characterize his work: the importance placed on statistics of an adequate
number of observations; and the minute, critical and accurate study of
rare individual events."[59] The second point was just as critical: the ability
to focus on the anomalous, odd event was one of the distinguishing features of great scientists—Rutherford's "genius to be astonished," or what
one insightful psychological study of scientists termed an "overalertness to
relatively unimportant or tangential aspects of problems," which leads sci-

entists to "look for and postulate significance in things which customarily would not be singled out."

There was, this study went on to observe, something bordering on the "autistic" or even "paranoid" in the grandiosity of this kind of thinking that frequently leads successful scientists to impute great meaning to the seemingly trivial. Indeed,

> were this thinking not in the framework of scientific work, it would be considered paranoid. In scientific work, creative thinking demands seeing things not seen previously, or in ways not previously imagined; and this necessitates jumping off from "normal" positions, and taking risks by departing from reality. The difference between the thinking of the paranoid patient and the scientist comes from the latter's ability and willingness to test out his fantasies or grandiose conceptualizations . . . and to give up those schemes that are shown not to be valid. It is specifically because science provides such a framework of rules and regulations to control and set bounds to paranoid thinking that a scientist can feel comfortable about taking such paranoid leaps.[60]

That was Blackett precisely. "In most of his undertakings Blackett combined versatility of imagination with tough scepticism. He was not easily convinced even by his own ideas," said Ivor Richards.[61] In examining the 700 cloud chamber plates, Blackett found 14 that were a bit odd. A competing scientist, Carl Anderson at Caltech, was on the same track; the price of Blackett's skepticism and thoroughness was that Anderson would just beat Blackett and Occhialini to publication with a brief announcement of a major new discovery. But Anderson's was quickly followed in February 1933 by Blackett and Occhialini's publication of a twenty-seven-page article meticulously describing their new method of "making particles of high energy take their own cloud photographs," combined with the far more convincing and decisive proof they had found for the existence of a bizarre new particle.[62] It was the first bit of antimatter ever discovered: the so-called positive electron, or positron.

The possibility that such a particle could exist had been raised a few years earlier by Paul Dirac, a shy, strange, and intense Cambridge theorist who had developed a mathematical formulation that combined quantum theory and relativity to explain the behavior of the electron. His equations did not actually predict the existence of an antielectron, but it was a curious feature of them that they worked just as well from a mathematical standpoint for such a hypothetical particle possessing a positive charge and negative energy. Blackett realized that the fourteen odd tracks were consistent with a particle having the mass of an electron. But under an applied magnetic field they curved in the opposite direction from most of the electron tracks—indicating that they were carrying a positive rather than a negative charge.

Blackett would later explain his delay in publication by observing, a bit tongue in cheek, that no one had taken Dirac's theory seriously at the time. But his own superabundance of caution was more to blame: determined to eliminate every other possible explanation, Blackett first carried out an exhaustive statistical analysis to establish how often positron-looking tracks might be produced by chance some other way. Finally satisfied, Blackett presented their findings at a meeting of the Royal Society on February 16, 1933. Unlike Anderson's paper, Blackett and Occhialini's explicitly made the connection to Dirac's theory and also provided solid evidence of electrons and positrons being created simultaneously. Over the next two days the *New York Times* ran several stories about their discovery, reporting from London that it was being hailed by physicists as "one of the most momentous of the century." Several nominating letters received by the Swedish Academy proposed that Anderson and Blackett share the Nobel Prize for their independent discovery of the positron. In the end, however, Anderson alone would receive the 1936 award for the milestone.[63]

With Blackett's growing renown came increasing friction with Rutherford. Zuckerman would say of Blackett that he was "brought up more to give orders than to seek counsel . . . a man to whom thought and action were the same." If anything, Blackett and Rutherford were too much alike in that way. Later that year the final eruption came between the two. Blackett emerged one day from Rutherford's office "white-faced with rage," recalled a colleague, and announced, "If physics laboratories have to be run dictatorially, I would rather be my own dictator." Another colleague thought that Blackett had also become fed up with the "still feudal-Victorian environment of Cambridge," which he found politically oppressive for a man of his increasingly left-wing views.[64] In the fall of 1933 he moved to London and a position at Birkbeck College. Recently integrated into the University of London, Birkbeck had originally been founded as the London Mechanics' Institute and retained its proletarian character, serving part-time and evening students who were pursuing a degree while working a job. Blackett took a flat in Gordon Square, in the heart of Bloomsbury, and immediately began laying plans to expand his cosmic ray studies with a new detector incorporating a huge, 11-ton magnet to be installed 100 feet belowground in an abandoned platform of the Holborn Underground station. The London tabloid press, never at a loss for a cliché, hailed him as a new "Sherlock Holmes," hunting beneath the streets of London for clues about the mysteries of the universe.[65]

4

Defiance and Defeatism

IF BLACKETT STOOD OUT from other left-wing intellectuals of the 1930s it was because he was no pacifist: for all of his political radicalization he always carried in his core a strand of the young naval cadet who had "enjoyed shooting at the enemy during the war." But to most of his scientific colleagues, opposition to war and to the exploitation of science for war were part and parcel of their growing political activism. In June 1934, 40 percent of the physicists at the Cavendish Laboratory signed a letter circulated by the Cambridge Scientists' Anti-War Group protesting the use of scientific research for military purposes.[1]

The group's founder was J. D. Bernal, who was carrying out pioneering work at the Cavendish in the new science of crystallography, using X-rays to explore the three-dimensional structure of complex biological molecules. He was also a communist so thoroughly committed to Soviet Marxism that he would defend not only Stalin's purges but even the pseudo-scientific claptrap of Trofim Lysenko, the uneducated peasant's son whose theories about plant breeding would become Soviet orthodoxy and lead directly to the persecution of a generation of Soviet geneticists who dared to point out the fallacies of Lysenko's assertions.

Bernal was, as well, a precocious social rebel of a type that would become tiresomely familiar in the 1960s, rationalizing a self-absorbed pursuit of sexual adventure as if he were somehow striking a blow for the liberation of mankind. Once, while his wife was six months pregnant, he managed to have affairs with three other women during a single two-week period. Neither his radical politics nor his personal morals won him many admirers in Cambridge.[2]

The CSAWG, on the other hand, had attracted a broad spectrum of support since its founding two years earlier. Disarmament, renunciation of war, and even outright pacifism had become not merely respectable; they were seen by many from across the British political spectrum as the only sane course.

"The virtues of disarmament were extolled in the House of Commons by all parties," wrote Winston Churchill in his account of the "locust years" leading up to the Second World War. He did not dwell on the merciless personal attacks he came under in the House, and in the press, for even suggesting there might be danger in a rearming Germany. The German delegation to the 1932 Geneva Disarmament Conference had demanded the removal of all remaining restrictions imposed by the Versailles treaty on the size of its military forces, and support in Britain for that step was considerable. *The Times*, the staid voice of the establishment, called for "the timely redress" of Germany's "grievances"; the left-wing *New Statesman*, now under the editorship of Blackett's old friend Kingsley Martin, endorsed "the unqualified recognition of the principle of the equality of states." Churchill, by then a backbencher who had split with his own Conservative Party over its policies on India, rose to attack the British government's proposal at Geneva that would slash France's army by 60 percent and allow Germany's to grow to the same size. Years later he vividly recalled the "look of pain and aversion" on the faces all around him when he was done. Members of all three parties, Tory, Liberal, and Labour, leapt to their feet to denounce him as "a disappointed office seeker," a man on a "personal vendetta" out to "poison and vitiate the atmosphere" of peace and cooperation built at Geneva. He was a "scaremonger" and a "warmonger."[3]

Even after the rise to power of Adolf Hitler in 1933, *The Times* quickly reassured the British public that all the "shouting and exaggeration" in Germany was "sheer revolutionary exuberance." So, too, did the British government, which went so far as to discredit the reports of its own diplomats throughout the spring of 1933 about the Nazi outrages taking place—attacks on synagogues, bonfires of books written by Jews, daily public beatings, clubbings, and even murders by Nazi storm troopers on the streets of Berlin—and sent conciliatory messages assuring the German government that no one in Britain believed such obviously exaggerated tales.[4]

Not even Hitler's subsequent open breach of the Versailles treaty shook the widespread belief that Britain needed to show the way to peace through its own example of disarmament. There was a lack of imagination in the British assumption that Germany must be like Britain in abhorring war. There was also a peculiarly British kind of self-admiration among the country's professional diplomats, but also among the larger body politic, that looked upon being calm and reasonable as the right way to handle unpleasantness. Hysteria was un-British; the French, with their emotional hatred for the Boche, were obviously overreacting to developments in Germany. The British foreign secretary, Sir Samuel Hoare, denounced "those people" who take "morbid delight in alarms and excursions"; Lord Lothian (who would be Britain's charming and down-to-earth ambassador in Washington during Britain's dark days in 1940) declared it "unpatriotic" for Britons to doubt Hitler's "sincerity" in wanting peace.[5]

The British left was instinctively antifascist but even its antifascism was no match for its pacifism, when it came to responding to Hitler's military buildup. "We on our side are for total disarmament because we are realists," the Labour leader Clement Attlee asserted. When the government proposed an extremely modest increase of the Royal Air Force, by three squadrons, Attlee replied, "The secretary of state for air is very carefully laying the foundation for future wars." In June 1935 the League of Nations Union announced the results of a nationwide survey. Eleven million Britons had been asked to subscribe to a "Peace Ballot" calling for disarmament, including the entire abolition of military air forces. More than 92 percent agreed.[6]

Churchill laid at least some of the blame for the embrace in Britain of such "defeatist doctrines" upon "the mood of unwarrantable self-abasement into which we have been cast by a powerful section of our own intellectuals."[7] He was right. It was an enduring peculiarity of memories of the Great War that they reversed the usual process of nostalgic amnesia; the passage of time had made the war more terrible, not less. The years 1929 and 1930 brought, in the words of Robert Wohl, a spate of "pessimistic, cynical, and sometimes very bitter and brutal" books and plays about the war.[8] The men in *All Quiet on the Western Front* (which sold 250,000 copies in its English translation the first year), *Good-bye to All That* (which made Robert Graves a small fortune), *Journey's End* (which became the smash hit of the London stage in 1929, running 594 performances at the Savoy), and dozens of similar works that came out that year bore no relation to conventional war heroes of literature. They die ridiculous, meaningless, gruesome deaths assaulting pointless objectives; honorable officers anesthetize themselves in rivers of alcohol; patriotism and heroism exist only to evoke horse laughs; even courage has no rational meaning since men are as likely to be killed no matter what they do.

A few perceptive critics noted that all of these portrayals of war were strongly shaped by a superimposed romantic narrative that elevated the experience of the individual soldier to the exclusion of any larger understanding of how or even why the war was fought. Though they purported to offer an unvarnished account of the "real war" in all its gritty horror and eschewing the sentimentality of clichés about valor and heroism, they were just as chockful of their own kinds of sentimentality and clichés of memory. ("It would be impossible to count the number of times," notes the critic Paul Fussell, that "the Valley of the Shadow of Death" and "the Slough of Despond" are invoked by these authors to describe the morasses of mud, ice, and bodies in the trenches.) That the Allies had won a series of smashing victories in the summer and autumn of 1918, in often brilliantly planned and commanded operations, or even the fact that the Allies had won the war, was lost in a personal narrative of gore and horror in the trenches.

One veteran vainly tried to point out that the war had not really seemed so bad at the time as it did now; nor was the almost universal trope of a halcyon and pastoral prewar England, contrasted with the hell of the front, even a very accurate picture. "One was not always attacking or under fire," he complained. "And one's friends were not always being killed. . . . And friendship was good in brief rests in some French village behind the line where it was sometimes spring, and there were still fruit trees to bloom, and young cornfields, and birds singing."[9] But, as Wohl observes, the literary version of the war easily won out:

> By the end of the 1920s, most English intellectuals believed that the war had been a general and unmitigated disaster, that England's victory was in reality a defeat, and hence that the men who had caused England to enter the war and to fight it through to the bloody end were either mercenary blackguards or blundering old fools. Such ideas could rally radicals as well as reactionaries.

There was another important political consequence to this line of reasoning: "In England the partisans of appeasement and peace at any cost found it useful to present themselves as the authentic representatives and heirs of the men who manned the trenches."[10] The Cambridge antiwar groups had gotten their start with a 1933 Armistice Day march to the town's war memorial to lay a wreath bearing the motto, "To the Victims of a War They Did Not Make, from Those Who Are Determined to Prevent All Similar Crimes of Imperialism." (Despite the bit about "Imperialism," the march attracted not just the hard left. There were also a significant number of Christian pacifists in the group.) By 1935 the Cambridge Scientists' Anti-War Group had graduated to staging protests against the RAF's annual air show at nearby Duxford airfield and issuing manifestos, frequently mailed to the prominent scientific journal *Nature*, signed by long lists of Cambridge scientists calling on their fellow "scientific workers" to refuse to lend their technical know-how to the cause of war. "The practical working of modern civilisation depends so largely on technical knowledge that if everyone with scientific training were to act for one common aim, that aim could be achieved," began one statement signed by Bernal and twenty-one other Cambridge scientists. "Probably the majority of scientific workers the world over prefer peace to war, and there can be no doubt that English scientific workers as a body are united in this matter," they declared.[11]

ADDED TO THE ACCEPTED HORRORS of the old war was a new horror, destruction from the air. During the 1930s, Harold Macmillan would recall many years later, "we thought of air warfare . . . rather as people think of nuclear warfare today."[12] The theories of Giulio Douhet, an Italian army officer whose ideas became well known when his book *The Command of the Air* was widely translated after his death in 1930, foresaw a not very distant future in which wars would be decided by the massive bombardment of civilian population centers by opposing air forces in the opening instant of a conflict. "Nothing man can do on the surface of the earth can interfere with a plane in flight, moving freely in the third dimension," wrote Douhet. "All the influences which had conditioned and characterized warfare from the beginning are powerless to affect aerial action."[13]

It was a theme widely picked up in newspaper articles and popular books as yet another reason why war had become unthinkable as a means of resolving differences between nations. "A bombardment with a mixture of thermite, high explosives and vesicants would kill large numbers outright, would lead to the cutting off of food and water supplies, would smash the system of sanitation and would result in general panic," wrote Aldous Huxley in the *Encyclopaedia of Pacifism*. "The chief function of the army would not be to fight an enemy, but to try to keep order among the panic-stricken population at home." The Cambridge Scientists' Anti-War Group carried out a number of rather amateurish experiments aimed at "proving" the impossibility of protecting the urban populace against gas attack.[14]

But, again, it was hardly just the left and committed pacifists that embraced this apocalyptic view of modern war. The General Strike of 1926 reinforced the view of Britain's ruling elite that modern industrial society was held together only by the slenderest of threads; the urban working class in particular could not be relied upon in a crisis. "Who does not know that if another great war comes our civilisation will fall with as great a crash as that of Rome?" asked Prime Minister Baldwin. But it was another observation of Baldwin's that would crystallize the nearly universal conventional wisdom for avoiding war, no matter what Hitler might do: "No power on earth can protect the man in the street from being bombed. Whatever people may tell him, the bomber will always get through."[15]

On August 8, 1934, amid a spate of articles about the terrors that awaited a nation plunged into modern aerial warfare, a letter appeared in *The Times* from Professor F. A. Lindemann of Oxford University:

Sir,—In the debate in the House of Commons on Monday on the proposed expansion of the Royal Air Force, it seemed to be taken for granted on all sides that there is and can be no defence against bombing aeroplanes and that we must rely entirely upon counter-attack and reprisals. That there is at present no means of preventing hostile bombers from depositing their load of explosives, incendiary materials, gases, or bacteria upon their objectives I believe to be true; that no method can be devised to safeguard great centres of population from such a fate appears to me profoundly improbable.

If no protective contrivance can be found and we are reduced to a policy of reprisals, the temptation to be "quickest on the draw" will be tremendous. It seems not too much to say that bombing aeroplanes in the hands of gangster Governments might jeopardize the whole future of our Western civilization.

To adopt a defeatist attitude in the face of such a threat is inexcusable until it has definitely been shown that all the resources of science and invention have been exhausted. The problem is far too important and too urgent to be left to the casual endeavours of individuals or departments.

The whole weight and influence of the Government should be thrown into the scale to endeavour to find a solution. All decent men and all honourable Governments are equally concerned to obtain security against attacks from the air, and to achieve it no effort and no sacrifice is too great.[16]

Lindemann was Oxford's professor of experimental philosophy and director of its Clarendon Laboratory, which as a center of physics research was but a distant rival to Rutherford's Cavendish Laboratory at Cambridge. But much more important, Lindemann was a member of Churchill's growing circle of official and unofficial contacts who shared his alarm over the increasing threat of Nazi Germany and who provided him with a steady stream of expert knowledge with which to challenge the government's policies.

It would be hard to imagine a man more unlike Churchill in his personal tastes. An ascetic bachelor, Lindemann was a strict vegetarian who did not smoke or drink. He had met Churchill when he partnered Mrs. Churchill at a charity tennis tournament, and the two men immediately hit it off despite their superficial differences. The son of a wealthy Alsatian who settled in England and took British citizenship, Lindemann was rich, very conservative, and in Zuckerman's words "enigmatic and 'grand.'" He reveled moving in important social and political circles. He had a solid reputation as a physicist, though had not done any significant original research of his own for some time. More exceptionally, he had conducted some astonishingly daring flying experiments during the First World War to develop a method for recovering from a tailspin. Rejected by the army for his German parentage, Lindemann had been taken on by the Royal Aircraft Factory at Farnborough to work on scientific problems related to flight; despite a vision defect he took flying lessons, and then personally made a series of flights in which he deliberately put his plane into a spin and coolly recorded what happened each time. From 1932 on he was a regular visitor of Churchill's, motoring his Rolls-Royce over from Oxford for weekends at Churchill's country estate, Chartwell, where the men would often talk into the small hours of the morning, discussing the gathering threats in the world. "Lindemann," said Churchill, "became my chief adviser on the scientific aspects of modern war."[17]

A few weeks after his letter to *The Times*, Lindemann and Churchill traveled to Aix-les-Bains, the French lakeside resort where Baldwin (then the Conservative leader in the coalition government) was vacationing, to press the idea for government action on a scientific study of air defense.

Lindemann followed up with two more letters to *The Times* answering a rejoinder from a pacifist who objected that improved air defenses would only make disarmament less urgent. "World opinion is definitely opposed to burglars," Lindemann replied, "yet Mr. Mander, I have no doubt, does not consider it superfluous to lock his front door at night."[18]

Churchill thought that the job should be given to a high-level body outside usual department channels and proposed establishing a special subcommittee of the Committee of Imperial Defence, the board that coordinated all of the military services. But in the meanwhile the Air Ministry's own scientific civil servants had been pushing the issue as well. The problem of defending population centers and other targets against enemy bombers was actually never quite so bleak as popular opinion held. It was true enough that since the end of the war the RAF had placed the greatest emphasis on building up its bomber force and that its overarching strategy was very much in line with Douhet's thinking about retaliation as the linchpin of air strategy. But it had not entirely neglected fighter defenses. Yearly war games that pitted the RAF's Air Defence of Great Britain Command against a mock force of attacking bombers had shown that the bomber did *not* always get through: in the 1932 exercise, 50 percent of the daytime raids and 25 percent of the nighttime raids were successfully intercepted. The real problem was timely and accurate early warning. During the Great War a system had been developed to relay observer reports along the coast via dedicated telephone lines to a command center in London, which in turn could dispatch fighters from airfields around the city. The trouble was that London's proximity to the coast left little time to react, and indeed nothing in England was more than seventy miles from the coast. The new, much faster bombers entering service cut the margin even closer.

In June 1934 the Air Ministry's director of scientific research, H. E. Wimperis, had asked his assistant A. P. Rowe to look into what was being done from a scientific standpoint to improve the air defense situation, and as Rowe recalled he reached exactly the same conclusion that Lindemann had—almost nothing:

> It was clear that the Air Staff had given conscientious thought and effort to the design of fighter aircraft, to methods of using them without early warning, and to balloon defences. It was also clear however that little or no effort had been made to call on science to find a way out. I therefore wrote a memorandum summarizing the unhappy position and . . . said that unless

science evolved some new way of aiding air defence, we were likely to lose the next war if it started within ten years.[19]

Even in the air force and navy, services that certainly by the 1930s depended upon scientific innovation for their very existence, the suspicion of science and scientists ran extraordinarily deep. England's scientists may or may not have been "united" in disdaining war work, as J. D. Bernal claimed, but uniformed officers were at least as united in disdaining civilian scientists, even the ones who did work for them. Rowe himself had received a vivid introduction to traditional military attitudes when he joined Wimperis's staff in the mid-1920s and tried to have a bomb tested in the wind tunnel at Farnborough, only to be told that the directorate of scientific research was to limit its inquiries to aeronautical questions; armaments were not its business. After a long bureaucratic struggle Wimperis was made an associate member of the interservice Ordnance Board, whose chairmanship rotated among an admiral, a general, and an air marshal. Rowe was sent as the scientific directorate's representative. Noticing that he was the sole civilian in the room, he took what he hoped was an inconspicuous spot at the table. It did not work. The chairman opened the meeting by announcing, "And today we have with us Mr. Rowe of the Air Ministry. Now perhaps, Mr. Rowe, you would like to let us know what you have to say." Rowe replied modestly that he had come to listen, not to speak. "In that case, Mr. Rowe," replied the chairman, "we need not detain you any longer."[20]

Now, ten years later, Wimperis took what Rowe rightly called "a bold step" for a civilian bureaucrat in the military services in this climate, and submitted directly to the Air Staff and the secretary of state for air, Lord Londonderry, a detailed and forceful proposal that a Committee for the Scientific Survey of Air Defence be established with a wide-ranging mandate "to consider how far recent advances in scientific and technical knowledge can be used to strengthen the present methods of defence against hostile aircraft." To ensure that this was more than just a rubber-stamp committee or a cover for business as usual, Wimperis urged that three outside experts— all distinguished scientists with military experience and unimpeachable integrity and independence—be brought in to serve on the committee. For the chairman he proposed Henry Tizard, a physical chemist at Oxford; the other two members would be A. V. Hill, a professor of biology at University College, London, and P. M. S. Blackett, "who was a Naval Officer before and during the War, and has since proved himself by his work at Cambridge as one of the best of the younger scientific leaders of the day." Rowe and Wimperis would make up the rest of the committee.[21]

"The five men who met under Tizard's chairmanship in Room 724 of the Air Ministry from 11 a.m. until 1.45 on the morning of 28 January 1935 were to help transform Britain's defences between 1935 and the outbreak of the war," states Tizard's biographer Ronald W. Clark.[22] They would also, it is no exaggeration to say, revolutionize the model of the application of scientific expertise to the art of war.

Tizard was a friend of Lindemann's—he had helped secure Lindemann's Oxford appointment—and like Lindemann had rather improbably learned to fly during the war to conduct scientific research for the air force. Tizard worked out a brilliant system that combined radio signals, sequential photography, and a pen recorder to precisely track the release and fall of a bomb from an airplane as part of the successful effort to develop the Royal Flying Corps' first practical bombsight. (Like Lindemann, too, he suffered from poor vision, the result in his case of a boyhood accident that left one eye damaged. He had been rejected for service by the navy for that reason. But he was accepted by the army, transferred to the Royal Flying Corps in 1915, and talked the authorities into letting him take flying lessons by proposing to fly only on days when the weather was too bad for the regular student pilots to go up.)

Tizard's ability to speak the language of airmen and understand their problems would carry weight. He also had a tough, skeptical, independent mind. As a student at Oxford he had learned an invaluable lesson about questioning received wisdom when his chemistry tutor had suggested he ought to do some serious reading during the long vacation:

Observing perhaps a lack of enthusiasm on my face, he turned to his shelves and picked out a new-looking book entitled *Chemical Statics and Dynamics*. "This is an important subject," he said, "but the book is rather too mathematical for me. I wish you'd take it away and see how many mistakes you can find in it." I believe this was the first time that any senior man had indicated that I might know more than he did about something that was worth knowing, and the first time that anyone had suggested to me that there might be mistakes in a printed book of science.[23]

Tizard quickly did find a mistake (which "aroused a detective spirit") and by the time he was done he had a long list of corrections, which he duly sent off to the author, receiving a gracious acknowledgment in return. During the war, after his work on the bombsight, he carried out a study of Allied bombing operations in France and concluded that there were so many sources of error, besides just aiming—target selection, navigation problems, poor visibility, difficulties in identifying the target—that the chance of a bomb ending up anyplace where it would have an actual effect on the enemy was "small," not exactly what his superior officers in the RFC wanted to hear.

The third outsider on the committee, A. V. Hill, was another man cut from the same straight-talking mold. He had shared the 1922 Nobel Prize in medicine for studies of muscle physiology. In the war, he had done groundbreaking work on statistical analysis of antiaircraft fire for the army.

In Wimperis's memorandum to Lord Londonderry he had urged that "no avenue, however seemingly fantastic, must be left unexplored" in the search for improved air defenses that could protect Britain's population centers. In preparation for the committee's first meeting Wimperis decided to try to get a definitive answer about one such fantastic avenue that had a long history. For years the Air Ministry had been pestered by inventors who promised that with sufficient time and money ("particularly money," Rowe noted) they could develop a "death ray." Hoping to get them out of their hair the ministry had offered £1,000 for the successful demonstration of such a device that could kill a sheep at 100 yards. Unfortunately this only encouraged more cranks when word got around about the promised cash prize. In the first week or two of January 1935, Wimperis phoned Robert Watson-Watt, who directed the National Physical Laboratory's Radio Research Section, and asked him if such a ray that could disable an aircraft or its crew was a physical possibility. Watson-Watt had his assistant A. F. Wilkins work out the numbers and it was immediately apparent that the amount of energy required was so vast as to be impractical. "Well, I wonder what else we can do to help them," Watson-Watt then said to Wilkins. The two talked over a few ideas and recalled reports that radio signals sometimes fluttered when an airplane flew nearby. Wilkins then did another calculation to see how much energy would be needed to bounce a radio signal off an airplane and receive an echo. The results were distinctly promising.[24]

Lindemann and Churchill meanwhile were still lobbying hard for a higher-level committee to be established and were suspicious when they learned the Air Ministry had beaten them to the draw in setting up its own panel. Churchill thought that entrusting the job to the same department that had failed so miserably to date was to guarantee just more inaction. In fact, the Tizard Committee moved with a speed that probably still holds an all-time record in the annals of military bureaucracy. At their first meeting Wimperis reported Watson-Watt's preliminary findings; at the second meeting, on February 21, they reviewed a full report that they asked Watson-Watt to prepare; and five days after that Watson-Watt, Wilkins, and Rowe were huddled before an oscilloscope watching the green trace rise and fall as a Heyford bomber flew through the radio signal from a nearby BBC short-wave transmitter. A week later Wimperis sent a memorandum to Air Marshal Hugh Dowding, the member of the Air Staff responsible for research and development, which began, "We now have in embryo a new and potent means of detecting the approach of hostile aircraft, one which will be independent of mist, cloud, fog, or nightfall."[25] Wimperis proposed that the Air Ministry provide £10,000 to continue the research. Dowding approved it on the spot. It was the beginning of the radar system that five years later would save the nation in the Battle of Britain, with Dowding himself in command as the commander-in-chief of Fighter Command.

Two months later, in April 1935, Churchill's political persistence paid off and his Air Defence Research Sub-Committee of the Committee of Imperial Defence was established, with himself as one of the members. Lindemann was added to the Tizard Committee. To confuse matters further, Tizard's committee was decreed to be simultaneously an Air Ministry committee and a technical sub-sub-committee of the Committee of Imperial Defence.

Churchill's enthusiasm for things scientific was genuine, if naive. He had a Victorian-era fascination with technology and statistics, which was fine; more troublesome, he had a weakness for the picture of the lone amateur inventor upending establishment thinking with a revolutionary idea that required a courageous champion to see it through. As first lord of the Admiralty he had played a real part in championing the development of the tank in 1915 against the resistance of the army ("Winston's Folly" it was called in Whitehall), which only confirmed his belief in his ability to pick a winning idea and become its patron. Soon Churchill and Lindemann were bombarding Tizard with all sorts of half-baked schemes, the loopiest of which were aerial mines that would intercept enemy bombers with a curtain of explosives drifting down on parachutes, and a device to create artificial updrafts to flip an airplane over in mid-flight. Another Lindemann-Churchill pet project was infrared detection of aircraft, which while not ridiculous in principle had been shown already to be extremely difficult in practice given the ease with which heat emissions from an aircraft engine could be shielded.

Tizard, Hill, and Blackett became increasingly incensed by Lindemann's efforts to push these hobby horses, especially since it would mean giving less attention to the crucial technology of radar. Blackett described how the issue came to a head the following year:

> It was not long before the meetings became long and controversial; the main point of dispute concerned the priorities for research and development which should be assigned to the various projects which were being fathered by the Committee. . . . On one occasion Lindemann became so fierce with Tizard that the secretaries had to be sent out of the committee room so as to keep the squabble as private as possible. In August 1936, soon after this meeting, A. V. Hill and I decided that the Committee could not function satisfactorily under such conditions; so we resigned.[26]

Lindemann's defenders would later assert that he and Churchill were always fully behind radar and were merely pressing for other promising ideas to receive attention as well. A thick file in the Air Ministry's archives labeled "Correspondence with W. S. Churchill, M. P. Air Defence Research Committee" suggests otherwise; it shows Churchill and Lindemann single-mindedly pursuing the aerial mine and infrared detection schemes and lambasting Tizard and the committee for their "half-hearted" and "slow-motion" efforts on these two projects. R. V. Jones, Lindemann's former student, recalled Lindemann mocking his rivals at this time with a parody of Omar Khayyám: "Something along the lines of 'The Blackett and the Tizard keep the courts where Trenchard once did sleep,'" Trenchard being the first chief marshal of the RAF. Following Hill's and Blackett's resignations, Lindemann smugly wrote the secretary of state for air that with those two obstacles out of the way, he was confident "this extremely important work will now be pursued more effectively and with due energy."[27]

Zuckerman observed that he could "not imagine any two people less likely to trust or like each other" than Blackett and Lindemann—Blackett "straightforward, leftish, Bohemian and unconventional," Lindemann high Tory, patrician, and calculating. The same went for Hill and Lindemann. And Tizard, though an old friend, found his patience wearing thin, too, especially when he discovered that Lindemann was repeatedly working through Churchill to revive in his Committee of Imperial Defence subcommittee ideas that had already been shot down on technical grounds by the Tizard Committee. "The trouble about Lindemann is that he is getting so infernally unscientific," Tizard wrote at one point. "He takes no trouble to examine even his own proposals properly, let alone anyone else's."[28]

Tizard, by contrast, saw that developing mutual trust with the air force's officers and researchers was key to getting anything done. Early on, Tizard admonished Lindemann that if he could not abide by the committee's decisions and understand the damage he was doing, he should not be a part of it. "What disturbed me in our conversation the other day," Tizard wrote him, "was your attitude that the people at the various Research Establishments were sure to be unhelpful, and slow to work out any ideas but their own." Just before the meeting that prompted Blackett's and Hill's resignations, Tizard again chastised Lindemann that "far from 'accelerating progress' you are retarding it. . . . If you want to have the work pressed with greater vigour, at the expense of other important work, you must surely start by convincing willing colleagues of the wisdom of doing this, instead of complaining to other people that they are slackers."[29]

For once Churchill was outmaneuvered in a skirmish of bureaucratic warfare. Following Blackett's and Hill's resignations the secretary of state for air promptly dissolved the committee, then reestablished it with all of its original members—save for Lindemann. The animosity between Churchill's favorite scientist and Blackett and Tizard would flare up again during the war when Churchill became prime minister and brought Lindemann back to Whitehall with him, with important and tragic consequences both personally and for crucial questions of war strategy.

AS THE AIR MINISTRY WAS taking its first steps to face the growing German air menace, "a most surprising act was committed by the British Government," as Churchill wrote with exaggerated restraint in his memoirs. On June 21, 1935, the first lord of the Admiralty informed Parliament that the British and German governments had reached an agreement on the future size of the two countries' naval forces. It was a stunning bow to Germany's unilateral repudiation of the military limitations imposed by the Versailles treaty. Even more stunning, Britain had not bothered to consult with or even inform her ally France of what it was about to do.

The Admiralty, in a lengthy memorandum enthusiastically backing the proposals that had been brought to London by Germany's ambassador, Joachim von Ribbentrop, argued that given the reality of Germany's undeniable intention and ability to rearm it behooved Britain to cut a deal quickly. Under the terms of the German proposal, Germany promised that her warship tonnage would never exceed 35 percent of Great Britain's. The British naval staff noted that "the German Government genuinely consider that they have made a generous and self-sacrificing decision, and that if the opportunity to close with the offer is lost, it is improbable that they will stop short at the 35 per cent level in building up their fleet." Moreover, "the statements of Herr Hitler . . . make it clear there is no prospect whatever of Germany coming to agreement on any question . . . except on the 35 percent. basis." Pressed by Ribbentrop for a quick answer—the German ambassador had expressed "disappointment that Herr Hitler's historic decision should have caused a delay"—the British government swiftly accepted.

On the question of submarines, the agreement astonishingly allowed Germany to build up to 45 percent of British strength—and, "in the event of a situation arising which in their opinion makes it necessary," 100 percent. (The only proviso was that "the matter shall be the subject of friendly discussion before the German Government exercise that right"; it was not, however, subject to British approval.) The British naval staff brushed off this as well, insisting that it really just came down to a matter of *Gleichberechtigung*, a favorite word of the Reich negotiators, which roughly translated as "equal rights":

> Should Germany exercise her power to build up to parity with ourselves in submarines, she could produce a formidable force of some 50 to 60 submarines (allowing for the fact that her first 12 are to be 250 tons each). This is a situation which must arouse some misgiving, but it is quite apparent from the attitude of the German representatives that it is a question of "Gleichberechtigung" which is really exercising their minds, and not the desire to acquire a large Submarine fleet. In the present mood of Germany, it seems probable that the surest way to persuade them to be moderate in their actual performance is to grant them every consideration in theory. In fact, they are more likely to build up to Submarine parity if we object to their theoretical right to do so, than if we agree that they have a moral justification.
>
> Apart from this psychological aspect of the question, the only other way to ensure a reasonable limitation of German submarine building is to keep our own tonnage as low as possible.

It is in any event satisfactory to know the limit beyond which the Germans do not intend to proceed. Under these circumstances, it is considered that the situation is acceptable.[30]

As part of the deal, Germany solemnly agreed to accede to the 1930 Submarine Protocol, a restatement of the principle that the laws of war applied to submarines just as they did to surface ships. In light of subsequent claims by German U-boat officers that the rules of international law permitted a submarine to torpedo merchant ships without warning when they were armed, or traveling in convoy, or with their lights darkened, or in war zones, or if they tried to radio for help, or if they otherwise placed the attacking U-boat in danger—assertions that even many historians who should know better have swallowed over the years—it is worth quoting the protocol in full:

The following are accepted as established rules of international law:

(1) In their action with regard to merchant ships, submarines must conform to the rules of international law to which surface vessels are subject.

(2) In particular, except in the case of persistent refusal to stop on being duly summoned, or of active resistance to visit or search, a warship, whether surface vessel or submarine, may not sink or render incapable of navigation a merchant vessel without having first placed passengers, crew and ship's papers in a place of safety. For this purpose the ship's boats are not regarded as a place of safety unless the safety of the passengers and crew is assured, in the existing sea and weather conditions, by the proximity of land, or the presence of another vessel which is in a position to take them on board.

The 1935 Anglo-German Naval Treaty was the apotheosis of the appeasement mind-set in Britain as Hitler was busily laying his plans for European conquest. Churchill called it the "acme of gullibility" to believe that the Germans would abide by their pledge to adhere to the Submarine Protocol once they possessed "a great fleet of U-boats," as they would now be permitted under the treaty. The truth was even worse than Churchill suspected. Hitler told Admiral Erich Raeder, the chief of the Marineleitung, the naval command, that he was overjoyed by this masterstroke and that the signing of the treaty was the happiest day of his life.[31] Construction of the Reich's new U-boat fleet was already in fact well under way. For over a year a steady stream of submarine components had been flowing from Germany to the Dutch front company IvS, then secretly reshipped by sea along the coast back to Germany for assembly. "The Führer demands complete secrecy on the construction of U-boats," Raeder recorded in his diary following a long discussion with Hitler in June 1934.[32] The naval staff had long settled on three designs for its new U-boat fleet: the Type II, a coastal vessel of 250 tons; the oceangoing Type VII of 625 tons and up; and a long-range Type IX of 1,000 tons. Now, just eleven days after the signing of the Anglo-German Naval Treaty, the first German U-boat since the war—a Type II boat designated *U-1*—was officially commissioned. Thirteen more would be completed by the end of the year. A modest stockpile of torpedoes was coming out of German factories as well.[33]

On September 28, 1935, Karl Dönitz took command of the First U-boat Flotilla. A week earlier the Reichsmarine ensign was replaced by the Nazi swastika throughout the fleet. Just over a year later Dönitz was promoted to Führer der U-boote, in command of the entire U-boat force. Now a Kapitän zur See, a full captain, he had continued to receive consistently excellent marks for his leadership, soldierly impression, popularity with officers and men, and energy. In his new task, he placed an unstinting emphasis on rigorous training and the inculcation of a "spirit of selfless mission-readiness" and an "attacking spirit" in his commanders and crews.[34]

THE NAZIS' INTENSIFYING PERSECUTION of Jewish scientists in Germany chipped away at both leftist pacifism and centrist apathy in the British scientific community. Blackett, for all of his continued entrancement with Marxist ideas about the role of science in advancing the working class, and his more than passing gullibility in accepting Soviet assertions about the USSR's unsurpassed support of scientific research, had parted company early with the pacifist party line of the British left on precisely this point. In December 1934, just before joining the Tizard Committee, he wrote a note on the state of science in Germany expressing concern that "many eminent scientists had left Germany, science was being used for anti–working class activities, and scientific fact was being deliberately distorted to accord with Nazi teachings."[35]

Some of the most eminent German Jewish scientists had departed even before the Nazi takeover, correctly reading the harbingers of rising anti-Semitism in Germany. The Hungarian-born aerodynamicist Theodore von

Kármán left for Caltech in 1930; Albert Einstein departed for Princeton in December 1932. Four months later Hitler promulgated the first anti-Semitic decree of the Nazi regime, the Law for the Restoration of the Professional Civil Service. It decreed that all "non-Aryan" employees of the state "must retire." Universities were state institutions, and some 1,600 scholars were immediately stripped of their posts, including a quarter of the nation's physicists. Among them were eleven past or future Nobel laureates. Hundreds more were solid, competent scientists but lacked the fame of an Einstein to be sought out by an American university eager to land a scientific star.

Blackett was active from the start in the effort to find places for the ousted scientists at English universities and laboratories, and he did what he could in his own lab. Otto Frisch, who would play a pivotal part in the design of the first atomic bomb at Los Alamos, was at the University of Hamburg when he and his fellow physicist Otto Stern were dismissed under the new law. Stern offered to make the rounds of the laboratories of colleagues he knew in France and England "and see if he could sell his Jewish collaborators," as Frisch remembered him wryly putting it. Stern told Frisch he would try to "sell" him to Madame Curie in Paris. "When he came back," Frisch recalled, "he said Madame Curie had not bought me, but Blackett had."[36]

Lindemann and A. V. Hill were also early participants in the effort to aid the refugees. Lindemann, though given to uttering occasional anti-Semitic remarks of the patrician British sort, immediately used his connections to procure a £20,000 grant from Imperial Chemical Industries to create positions for twenty displaced physicists and chemists from Germany. Hill, who became secretary of the Royal Society in 1935, had like Rutherford previously maintained that scientists should remain aloof from politics "as a condition of complete intellectual honesty." He revised that opinion in the face of the Nazis' outrages against what he termed "freedom of thought and research" and became a prime mover in establishing a more formal organization to aid the displaced scientists. With Rutherford lending his name as president and Hill as vice president, the Academic Assistance Council (later the Society for the Protection of Science and Learning) went to work raising funds and began providing modest grants, £182 a year to unmarried scientists, £250 for those with families, so they could be given places at British universities and other institutions. By the time the war began, the organization had raised £100,000 and aided some 900 scientists.

It was probably the single most powerful influence in awakening British scientific opinion in the late 1930s to the growing threat, and evil, of Nazism.

The same, though, could hardly be said for British opinion as a whole. After receiving a barrage of anti-Semitic hate mail, the council played down the fact that most of the scientists it was aiding were Jews. For fear of stirring up a further anti-Semitic backlash in Britain, the Royal Society, as a condition for offering office space and support, even vetoed the council's proposal to appoint a Jewish scientist to its board. Still, it was something. "More unanimously than any other intellectual group," C. P. Snow thought, "the scientists were anti-Nazi." Blackett's prominent role in aiding refugee scientists, as well as his example in rejecting the left's moral strictures upon scientists becoming involved in military work, was a powerful influence during this period. Blackett, in Snow's view, "was the chosen symbol of scientists on the Left. In fact, he spoke for the younger generation of scientists in the thirties very much as Rutherford spoke for the older."[37]

The hard left among the British scientists remained as doctrinaire, obdurate, and self-regarding as ever, and J. D. Bernal in particular was conspicuous in his absence from the serious work of aiding the Nazis' scientific victims. But by 1938 even he had done an about-face on the question of whether scientists could in all conscience help their government win a war against fascism. This, to be sure, may have had more to do with the Marxist dialectic than the facts about Hitler and his intentions. Bernal was writing a book titled *The Social Function of Science* that sang a Marxist paean to the Soviet regimentation of science for social progress; it was full of what Leonard Woolf once aptly termed the "slants, snides, sneers, and smears which Communists and Fellow Travellers habitually employ as means for building a perfect society."[38] Bernal wrote that it remained doubtful whether the armed forces of Great Britain would ever be employed in behalf of "the principles of democracy and civilization"; that the application of science to war was to most younger scientists "the worst prostitution of their profession"; nonetheless the one positive benefit of the assimilation of science into the modern war machine ("more than anything else") was that it has "made scientists look beyond the field of their own inquiries . . . to the social uses to which their discoveries are put."[39]

It was an infuriating enough argument—in effect, the real reason to fight Hitler was because it would hasten the longed for collectivization of science—that it provoked John Baker, an Oxford zoologist, to a furious and swift rejoinder. "Bernalism," he wrote in the *New Statesman*, "is the doctrine of those who profess that the only proper objects of scientific research are to feed people and protect them from the elements, that research workers should be organized in gangs and told what to discover, and that the pursuit of knowledge for its own sake has the same value as the solution of crossword puzzles."[40]

Be that as it may, Bernal in late 1938 produced a memorandum entitled "Science and National Defence," which he sent to B. H. Liddell Hart, the military correspondent of *The Times*. Almost exactly reversing his previous stance, this new manifesto called for the full wartime mobilization of British scientists. Bernal wrote an editorial for *Nature* advocating the same policy. The Association of Scientific Workers, a white-collar trade union dominated by Marxists who improbably tried to identify university professors and Ph.D. scientists as members of the working class (Bernal had helped revitalize the group in the 1930s and Blackett would assume its vice presidency in 1939 and its presidency in 1943), likewise dropped its previous opposition to the application of science to war. The association adopted a resolution in November 1938 stating, "While we regard war as the supreme perversion of science, we regard anti-democratic movements as a threat to the very existence of science. . . . We draw attention to the fact that the most efficient utilisation of science in time of emergency necessitates in time of peace a much wider application of science to all productive forces and social services."[41] The Royal Society added its support for the compilation of a Central Register of British scientists who could be called upon to assist the war effort, a proposal quickly accepted by the government; some 7,000 scientists were enrolled, including 1,175 physicists.[42]

THE DEVELOPMENT OF Britain's radar defense system was meanwhile rapidly forging deeper ties between the scientists and the military. In the summer of 1936 Tizard began pressing the air force to conduct realistic trials of the entire fighter defense system to see how it would work in practice. He was acutely aware that the early warning provided by radar was only as good as the system for relaying, synthesizing, and acting on that information. The air force reluctantly agreed to the tests, even though they were a huge drain on manpower, equipment, and fuel at a time when there was little of any of those to spare.

The trials, based at Fighter Command's Biggin Hill airfield just southeast of London, immediately showed that the procedures for filtering out contradictory data, funneling reports to the relevant operations centers, plotting the tracks of enemy bombers, allocating targets to various fighter units, calculating the courses the fighters should be instructed to fly to intercept were—in Tizard's words—"quite hopeless."[43] The 1936 tests were supposed to last a few months but instead went on for nearly two years. In the spring of 1938 more radar stations were coming online in the chain of stations being erected along Britain's southern and eastern coasts, and the need for additional scientists who could help train operators and work out the kinks in the equipment was becoming all too apparent. Tizard invited John Cockcroft, a leading member of the Cavendish staff, to lunch at the Athenaeum Club in London and explained the situation. Cockcroft promptly agreed to help, and eighty physicists from the universities of Cambridge, Oxford, Manchester, Birmingham, and London, duly sworn to secrecy but unpaid, were enlisted to spend the long vacation in the summer of 1939 at the air force's Bawdsey Research Station learning about radar.

Bawdsey, on the Suffolk coast near Felixstowe, was one of a number of bizarrely ornate nineteenth-century manor houses conscripted by the military for use in the war. Its grounds were the site of one of the radar stations of the Chain Home network, and it was chosen to be headquarters of the main radar research unit as well; Watson-Watt had been brought in as superintendent and A. P. Rowe as assistant superintendent. To both Tizard and Rowe it was manifest that the work on improving operational procedures of the radar defense system that had begun in the Biggin Hill trials needed to continue. Rowe assigned a small team under Eric C. Williams, who had been the first of the young scientists from a university to join the staff at Bawdsey, to analyze and compare the performance of the several control stations that had been established to see if he could determine what accounted for the differences in their ability to successfully intercept targets.[44]

This was something entirely new for scientists working for the military; as an "indirect but very real achievement" of the Tizard Committee, this new form of scientific-military cooperation would in retrospect rank in Blackett's view as every bit as important as the invention of radar itself. Blackett had moved from London to the University of Manchester in 1937, taking his 11-ton magnet with him, but continued to play an active part in defense affairs through the Tizard Committee, which had now taken on a larger responsibility within the Air Ministry as reflected in its new name, the Committee for the Scientific Survey of Air Warfare. The Biggin Hill study, Blackett explained, was the first official recognition that the actual operations of modern war are so complicated and change so fast that the traditional training of the serving officers and personnel is inadequate. In fact, many of the operational problems which arise when new equipment comes into service require for their solution the aptitudes of the scientific research worker: for he is trained to apply scientific methods to elucidate hitherto unknown and complex phenomena.

The traditional military view was that the scientists' role was to develop "weapons and gadgets," hand them over, and that was that. But now scientists were intimately involved in what previously had been the exclusive purview of military commanders: the running of operations.

The mutual confidence forged in the process broke down barriers on both sides; "so that," wrote Blackett, "when need arose, and it did arise very soon, many of the best academic research workers flocked out of the universities into radar stations, and later into service experimental establishments, where they became a vital part of the brilliantly creative, and sometimes obstreperous, teams, whose work had so profound an effect on the waging of the war." He continued:

> These developments implied a great measure of mutual trust and understanding between the senior service officers and the often brash and initially very ignorant scientists—ignorant, that is, of most things that went on outside a university research department. . . . Hitler's Third Reich saw no such collaboration. No doubt the almost unbroken German military successes of the first war years confirmed the highly competent military staffs in the view that they had no need to seek help from outside scientists, however brilliant. When the tide of war swept against Germany it was too late. Luckily for the Allies, Germany never produced its Tizard.[45]

Rowe, sometime in 1937, came up with a name for this new kind of scientific investigation into the business of running a war: "operational research." Williams's group was entered on the organizational chart at Bawdsey as the "Operational Research Section."[46]

IN THE SUMMER OF 1939 the final stations of the radar network were completed to form an unbroken sentry line, a chain of spidery steel towers standing guard against the nightmare threat of Britain's annihilation from the air by the Nazi war machine. Of the RAF's thirty-five fighter squadrons available for the defense of the island, twenty-nine were now equipped with the new fast monoplane Spitfires and Hurricanes in place of the biplanes that had filled their ranks just a few years earlier. Blackett left for a vacation, renting a cottage for the first of many summers he would spend in a beautiful spot in the Welsh mountains where many other leftish intellectuals had vacation homes, Bertrand Russell the most notable. "In our Welsh cottages," the Marxist historian Eric Hobsbawm wrote many years later with a certain good-natured self-mockery, "we voluntarily lived under the sort of conditions we condemned capitalism for imposing on its exploited toilers." ("How Left-Wing Was My Valley" was the title of his reminiscence.)[47] In August, Blackett wrote from Wales to a scientific friend, Michael Polanyi, admitting he was stunned by the news that Stalin had just signed a non-aggression pact with Hitler—adding that Polanyi, who had never shared Blackett's starry-eyed admiration for the Soviets, no doubt felt a certain degree of "Schadenfreude" over the anguish this act of Soviet expediency was causing the British left.[48]

By then appeasement was dead beyond all possible doubt, trampled by the long trail of German perfidy that reached its end with Hitler's betrayal of his pledge given at Munich, in September 1938, that the absorption of German-speaking regions of Czechoslovakia was his "last territorial claim" upon Europe. Five months later, Hitler sent his troops to seize the rest of Czechoslovakia and Prime Minister Neville Chamberlain announced that Britain was prepared to guarantee Poland's security.

In response, Hitler denounced the Anglo-German Naval Treaty. "The basis for it has been removed," he declared on April 28, 1939, in a long sarcastic speech to a snickering Reichstag in which he ridiculed a message from President Franklin D. Roosevelt seeking Germany's assurances of her peaceful intentions toward her neighbors. But Hitler solemnly disavowed any German plans to attack Poland (such reports, he said, were "mere inventions of the international press").[49]

The British prime minister had been genuinely shocked by Hitler's betrayal of the Munich pact, and took it personally. "In spite of the hardness and ruthlessness I thought I saw in his face," Chamberlain said, "I got the impression that here was a man who could be relied upon when he had given his word."[50]

5

Remedial Education

ON THE EVE OF WAR Britain's naval experts remained serenely confident that the U-boat menace had been vanquished for good. "The submarine," a 1937 British naval staff report concluded, "should never again be able to present us with the problem we were faced with in 1917."[1] Such confidence was the chief reason the Admiralty had been so untroubled at the prospect of Germany's acquiring the force of fifty to sixty U-boats permitted her under the Anglo-German Naval Treaty.

There were two main arguments that led to this reassuring conclusion. One was convoys: however reluctant the Admiralty had been in first adopting the practice, their effectiveness in 1917 and 1918 had been so decisive as to remove all doubt. The naval staff had drawn up extensive contingency plans for taking over the control and movement of all merchant shipping should war come again, and by the summer of 1939 a Shipping Defence Advisory Committee had been meeting monthly with shipowners to work out the practical details.[2]

The other source of British confidence was asdic, or sonar as it was now known in America. British, French, and American scientists had all been working on the idea at the time of the Armistice. Hydrophones, underwater microphones that passively detected sounds beneath the surface, had been instrumental in only four kills of German submarines during the war, but asdic promised much better results. Like radar, asdic worked by transmitting pulses of energy—high-pitched sound waves in the case of asdic—and then recording the time it took for an echo to return from a target; the longer the time, the greater its distance.

Had the war lasted another six months the system would have been ready for operational deployment on Royal Navy warships. The development of asdic had incidentally provided another example for Churchill's belief that military heads required substantial knocking to get them to accept new ideas. The work had been shepherded by a board of civilian scientists set up by the Admiralty in July 1915—this was the group that Rutherford had been brought in to help with—and was headed by Churchill's outspoken and like-minded ally Jacky Fisher. It was officially known as the Board of Invention and Research; Fisher's many enemies among the regular navy's tradition-minded officers called it the Board of Intrigue and Revenge. It earned little but suspicion and hostility at first. Nonetheless, the researchers made substantial progress in working out both the basic physical facts of undersea acoustics and the technology of a practical system for active detection of submarines by sound waves, and just before the end of the war the first prototype asdic set was fitted on a British research vessel for sea trials.

By 1921 senior naval officers were so enthusiastic about the new invention that they were convinced it had upended naval strategy altogether, pronouncing it an "epoch-making achievement" that would allow escorts to protect convoys on long passages through hostile waters and even permit the Royal Navy's mighty surface ships to "keep to the sea as in the golden days," rather than cowering in protected bases or skulking along behind a sprawling defensive screen of destroyers when they did sally forth. The standard asdic sets installed on British warships between the wars had a detection range of up to 5,000 yards and were equipped with a range recorder that used chemically treated paper, much like that of early facsimile machines, to draw a plot tracing the movement of a detected submarine.

British officials were divided over whether to make this new secret weapon known to rival powers. Political leaders thought disclosure would help peace efforts by convincing other navies that submarines were obsolete, and thus to accept British proposals to restrict their numbers if not abolish them outright. The Admiralty's director of intelligence pointed out the logical inconsistency in this thinking: either submarines were a threat to Britain's great surface fleets or they were not, and if the government truly believed that asdic had rendered the threat impotent then there was no need to care whether other naval powers kept building submarines; indeed, "if this country really had an antidote then we would let other nations waste their resources on submarines." In the end the policy of keeping asdic under wraps as long as possible prevailed; the very word remained an official secret until 1929.[3]

That did not, however, prevent the German navy from learning of the technology, and Dönitz more cannily played the game in reverse. "According to the English Press, England apparently believes herself equal to the U-boat danger on the grounds of her detection apparatus," he observed in early 1939. "Our goal must be *under all circumstances to leave England in this belief.*"[4] In any event, the British naval staff was far more worried about Hitler's newly constructed fleet of surface raiders, built around the two fast "pocket" battleships *Deutschland* and *Admiral Graf Spee.* These were warships of a new and intimidating design, powered by diesel engines that delivered a formidable cruising range of 21,500 miles and a speed of 28 knots, armed with six 11-inch guns, and they consumed an inordinate share of the British navy's planning and attention. In the judgment of the Royal Navy's official historian Stephen Roskill, it was probably the greatest mistake of the war that the Admiralty made.

IN TRUTH, Hitler's senior naval officers were deeply worried themselves.

On August 15, 1939, Dönitz received a message that a "submarine officers' reunion" would be held four days later. It was a coded instruction for his U-boat force to put to sea at once and take up war stations around England. Two weeks later, in the early morning hours of Sunday, August 27, Hitler held a bizarre conversation at the Reich Chancellery with a Swedish businessman, Birger Dahlerus. A friend of the Luftwaffe chief Hermann Göring, Dahlerus had become an unofficial emissary between Germany and Britain in the final days leading to Germany's invasion of Poland. Shuttling back and forth between Berlin and London, he saw himself as someone who understood both sides and could avert war where the professional diplomats had failed. Dahlerus had lived in England, and tried to convince Göring that Britain would this time stand by her commitments. Likely sincere though hopelessly naive (he would later testify as a character witness for Göring at Nuremberg), Dahlerus was in reality little better than a dupe of the Nazi government's effort to drive a wedge between Britain and Poland.

This time he came bearing a letter from Lord Halifax, the British foreign secretary and the man in Chamberlain's government whom Hitler had quite accurately picked out as the most likely to still be eager to grasp a "peaceful solution"—meaning a sellout of Poland that once again would let Hitler gobble up a neighboring territory without having to fight for it. Hitler, awakened by Göring at midnight to receive the self-appointed envoy, proceeded to deliver a twenty-minute harangue, pacing the floor, testifying to his sincere efforts to come to an understanding with the British and their incomprehensible refusal to grasp his proffered hand of friendship. Then, suddenly, the Führer stopped in the middle of the room and stood there staring, "his behavior that of a completely abnormal person," Dahlerus thought. Hitler started almost chanting an incantation of what he would do if there should be a war: *"Ich U-boote bauen, U-boote bauen, U-boote, U-boote, U-boote, U-boote!"* ("I will build U-boats, build U-boats, U-boats, U-boats, U-boats, U-boats!") Dahlerus glanced nervously at Göring to see how he was reacting to this display of the Führer's phantasmic temper, "but he did not turn a hair."[5]

Despite these threats, Hitler was profoundly equivocal about the U-boat arm. He had early on secured the loyalty of the Kriegsmarine's officer corps with the lavish promise of an expansive naval construction program, and the navy quickly became the most ardently pro-Nazi of the military services in Germany. Under the latest construction program—the Ziel ("Target") Plan, or Z Plan, which Hitler approved in January 1939—the navy was to grow by 1945 to a powerful fleet of 10 battleships, 4 aircraft carriers, 3 battle cruisers, 3 pocket battleships, 5 heavy cruisers, 48 light cruisers, and 68 destroyers.[6] It was a force that would, as in the last war, directly seek to challenge the Royal Navy's command of the North Sea, and even waters beyond. Both Raeder and Hitler rejected an alternative proposal to concentrate on quickly building up a large force of raiders, both U-boats and pocket battleships, to strike at British commerce.

The emphasis on conventional, big-ship sea power was the dominant German naval policy throughout the Nazi buildup, so much so that Dönitz had at first despaired upon being assigned to the U-boat force again: "I saw myself pushed into a siding," he said, when he learned in July 1934 that he was being moved from command of the cruiser *Emden* to the submarine arm.[7] Although the Z Plan also called for 249 submarines, the reality on the eve of the war was far short of anything approaching that figure. There were a total of only 57 U-boats in service on September 1, 1939, only 26 of those the oceangoing types capable of operating in the open waters of the Atlantic.

The outbreak of war found the rest of the German navy equally short of the ambitious goals of the Z Plan, with only 2 battle cruisers, 3 pocket battleships, 3 heavy and 5 light cruisers, and no main battleships or aircraft carriers at all completed. A gloomy assessment by Raeder noted that the navy "was in no way very adequately equipped for the great struggle with Great Britain"—a war, he pointedly added, "which according to the Führer's previous assertions, we had no need to expect before about 1944." In the short time since 1935, the submarine arm had become a "well-trained and suitably organized" force. But, Raeder continued,

> The submarine is still much too weak, however, to have any decisive effect on the war. The surface forces, moreover, are so inferior in number and strength to those of the British Fleet that, even at full strength, they can do no more than show that they know how to die gallantly and thus are willing to create the foundations for later reconstruction.[8]

Dönitz concurred, adding his own pessimistic appraisal. "We cannot expect the number of U-boats now on operation to be more than a petty annoyance to British commerce," he wrote. Of the 26 boats suitable for operation in the Atlantic, 3 were in the Baltic and 5 were not yet ready for active duty or were undergoing trials, leaving a total of 18 that had been able to take up stations for the fight against Britain as of September 1. He had sent every available boat to sea to be in a position to deliver the strongest possible opening shot at the onset of hostilities, but in a sustained fight the number on station at any time would quickly fall to about one third of the total available force, to allow for repair and resupply. Even the currently planned construction of new U-boats over the next six years would yield a force of less than half the number of operational boats that Dönitz calculated would be necessary to achieve a decisive result against the enemy's merchant shipping.[9]

That magic number was 300, Dönitz said, and Germany's failure ever to produce an operational fleet of oceangoing U-boats of that size would subsequently be cited by Dönitz himself as the chief reason he lost the Battle of the Atlantic. Many historians since have bought that assertion; scarcely a book written about the U-boat war does not cite the figure of 300 boats as the force Germany required. In fact, the number was almost purely arbitrary, based on no real analysis at all; Dönitz probably chose it as the largest figure he thought he could sell to Raeder and Hitler at the time. Commanders from time immemorial have sought to cover in advance the possibility of their failure by complaining of inadequacies in the men, arms, and supplies provided them. And Dönitz was if anything more optimistic than his superiors on the naval staff, some of whom argued that it would be irresponsible to send the U-boats to sea at all against such one-sided odds. Dönitz countered in memorandum after memorandum that he was confident the U-boats could hold their own.

Some of his confidence—though surprisingly little—stemmed from improvements in U-boat designs and technology that had taken place since the end of the last war. The new boats were certainly more solidly constructed, with welded components replacing rivets. The horsepower of their engines had doubled, increasing the top surface speed of the ocean-going boats from about 13 knots to 17 or 18. A system allowing both propeller shafts to be driven by one of the two engines at a time when cruising at low speed allowed for much more efficient fuel use.[10] But overall the design of the boats had changed remarkably little. The basic shape, functional layout, and sea performance of the U-boats that entered the war in 1939 were hardly different from those of the fleet surrendered at Harwich at the end of the last war.

Potentially far more significant were innovations in torpedoes. The new Type G7e was a masterpiece of German engineering. It was powered by a quiet electric motor running off batteries and gave off none of the telltale stream of air bubbles of the standard models, which were propelled by steam engines driven by an expanding vapor mixture of seawater injected into burning air and gas. The German U-boat designers had also invented a closed system that fired the torpedo without sending a blast of compressed air into the water with it; the new design used an air-driven piston to push the seawater and torpedo out of its tube.

The real change, though, was the tactics Dönitz had been honing to counter both convoys and asdic. Asdic could only detect submerged objects. Dönitz was contemplating night surfaced attacks, relying on the concealment of darkness and the U-boat's low silhouette to get within torpedo range of his targets. That would render asdic simply irrelevant. As for convoys, they were effective for a simple reason of arithmetic: it is not thirty times easier to find thirty ships together than it is to find one ship. If a U-boat did chance upon a convoy it usually could fire only one or two torpedoes at most before fleeing. But if the attack could somehow be continued to sink a significant percentage of ships in a convoy once it was located, the advantage of traveling in a convoy in the first place would be negated. Dönitz had worked on this problem and was convinced he had the solution: a concentration of targets would be met by a concentration of attackers. He devised a plan for deploying U-boats in loose groups along the Atlantic sea-lanes; when a U-boat spotted an enemy merchantman it would transmit a report by radio to the other boats in his group and to U-boat headquarters, which would in turn order additional groups to the area. Meanwhile, the first U-boat would continue to shadow and keep touch with the convoy just at the limit of visibility until nightfall, when the assembled group would move in for the kill. Dönitz tried out the concept in a naval staff war game in early 1939 and in exercises at sea that spring in the Bay of Biscay, and however contrived the conditions of the experiments he was sure they had proved his ideas. "The simple principle of fighting a convoy of several steamers with several U-boats as well, is correct," he wrote after the exercise.[11]

The key to making it work was a secure communication system linking the boats at sea with BdU—Befehlshaber der U-Boote, the commander of U-boats. The German military had such a system: since the 1920s it had been using a modified version of a commercial device called the Enigma to encode its radio traffic. The Enigma was a small box containing a typewriter keyboard and a set of lamps labeled with the letters of the alphabet. Pressing a key completed a circuit that sent an electric signal on a meandering path through three wired rotors to light up a corresponding letter; the rotors advanced their position with each successive keystroke, cycling roughly like a mechanical odometer, so that a new scrambler pattern was employed for every successive letter of the message. The radio operator wrote down the coded version of his message produced by the Enigma, then transmitted it by hand over his radio in Morse code. The operator at the other end reversed the procedure, typing the received message into his Enigma machine to rescramble it back to the original text.

Even if a machine fell into enemy hands the system retained its frightful impenetrability. By choosing a different starting position of the rotors, re-arranging the left-to-right order of the three rotors as they were inserted into the machine, and, in later models, by plugging a set of jumper cables into various jacks, literally trillions of different encipherment sequences could be generated. The German navy, since May 1, 1937, employed a particularly secure procedure for letting the intended recipient know which of those trillions of settings had been used for a particular message; it involved sending at the start of the message a separately encoded eight-letter code group that could only be deciphered using a printed table distributed once a month to users of the system. These external code tables added an extra layer of security to the navy's Enigma codes that made it all but impervious to mathematical methods of attack by any would-be code breakers.

Overall, Dönitz was supremely confident that he had in his U-boats a weapon that could not only win battles, but win the war. "It is clear that the attack on English sea-communications alone," he wrote in a memorandum to the naval staff on May 23, 1939, "can have war-decisive effect in a naval war against England."[12] Dönitz calculated that if he could sink 700,000 tons of shipping a month, he could destroy the merchant ships that Britain depended upon for her survival faster than her shipyards could replace them.

WITH THE COMING of the war that Winston Churchill had long warned of, Chamberlain could no longer keep out of the government the one man who had so forcefully stood up to Hitler through all the years of appeasement. On September 3, 1939, two days after the German invasion of Poland and the day war was officially declared between Britain and Germany, Churchill joined the cabinet as first lord of the Admiralty. It was the same post he had held a quarter of a century earlier, during the last war with Germany. In his first major speech to Parliament in his new position a few weeks later, Churchill began by saying how strange it was to be back in the same room at the Admiralty, in front of the same maps, facing the same enemy, dealing with the same problems. Then his face broke into an enormous grin. Glancing down at the prime minister, who was seated next to him, he observed in a tone full of innocent wonder, "I have no conception how this curious change in my fortunes occurred." The entire house roared with laughter—all except Chamberlain, who "just looked sulky," thought Harold Nicolson, an MP and author who kept a sharply observed diary.

"One could feel the spirit of the House rising with every word" that Winston spoke, Nicolson wrote.[13] Churchill's supreme confidence in the darkest hour, his ready wit, his boyish enthusiasm had already made him the hero of the moment. Congratulatory letters and telegrams poured into the Admiralty. All more or less said the same thing: Thank God you're back. Churchill swept into his old offices like a whirlwind, ordering up maps and charts, having the old octagonal table he remembered hunted up and restored to its place, issuing a veritable gale of short memoranda that first day and continuing at the same unbroken pace for the next nine months:

"Kindly let me know . . . ," "Please let me have . . . ," "What are we doing about . . . ," "Why is work being held up on . . . ," "Let me have on a single sheet of paper . . . ," and, with his inimitable love of archaic phrasing, "Pray consider this matter . . . ," "Pray get this on the move . . . ," "Pray meet together and put this into shape . . ." The "First Lord's Prayers," they were called. He had little red gummed labels printed up with the words ACTION THIS DAY, which he affixed to memoranda demanding immediate attention. One of his first instructions was that German submarines were henceforth to be referred to in all official papers and communiqués only as "U-boats."[14] He explained that this was to "avoid confusion" with references to friendly submarine forces, but with his instinctive ear for language he knew that calling them U-boats made them sound sinister, and different, from anything that other navies of the world possessed.

Bad news did not wait for the first lord to settle in. At 8:59 p.m. on that very first day of the war, the Malin Head radio station on the northwest coast of Ireland picked up a distress call from SS *Athenia*, a passenger steamer of the Donaldson Line bound for Montreal from Glasgow. The message began with the staccato alarm . . . . , Morse code for *SSS*, the signal for a submarine attack. Survivors later reported that after sending a torpedo crashing squarely amidships, blowing away the bulkhead between the boiler and engine rooms and destroying the stairs leading to the upper decks from the third-class and tourist-class dining rooms, the submarine had surfaced a bare 800 yards off the liner's port side. As passengers and crew were taking to the lifeboats, the U-boat fired one or two shells from its deck gun before vanishing into the night. Two British destroyers, a Norwegian cargo ship, and a private Swedish steam yacht picked up the survivors from the lifeboats, but 112 of the 1,418 aboard—including 30 Americans—died, killed by the initial torpedo blast, trapped belowdecks, or drowned in the chaotic rescue operation. Three quarters of the dead were women and children.

It had apparently been more a case of mistaken overeagerness than deliberate ruthlessness. Expecting that Britain and France might yet abandon the war once his swift conquest of Poland was a fait accompli, Hitler had ordered the U-boats initially to stick to the prize rules, and Dönitz's war orders so instructed. The next day Dönitz reiterated the orders with a signal to all U-boats: "By order of the Führer, on no account are operations to be carried out against passenger steamers, even when under escort." At a conference with Raeder on September 7, Hitler pointed to "the political and military restraint showed by France and the still hesitant conduct of British warfare." France had done nothing with its mighty army, which remained hunkered behind the defense of the Maginot Line on Germany's western border as Hitler's panzers plunged eastward through Poland. Raeder recorded the Führer's general policy instructions at the meeting: "Exercise restraint until the political situation in the West has become clearer, which will take about a week."[15]

The parallels with the *Lusitania* were all too apparent, and repercussions in America were also something Hitler wanted to avoid. The German government at first announced that the *Athenia* must have either been torpedoed in error by a British warship or struck a British floating mine. A far more audacious lie followed once Dönitz, Raeder, and Hitler ascertained the truth a few days later: that the attack had indeed been carried out by a German submarine that had just returned to port—*U-30*, under the command of Oberleutnant Fritz-Julius Lemp. Hitler's propaganda chief, Dr. Joseph Goebbels, took to the airwaves himself to reveal that the sinking had been part of a "devilish game" by the "British Ministry of Lies" to generate "a new *Lusitania* case": she had in fact been blown up by a bomb secretly placed aboard her by the British Admiralty. "We believe the present chief of the British navy, Churchill, capable even of that crime," Goebbels solemnly averred. Other German news reports pointed to photographs of British destroyers coming to the aid of the stricken liner as confirmation of this conspiratorial explanation: why else had British warships been so conveniently close by just at the right moment? Dönitz obediently fell in line with the cynical cover story. He personally swore Lemp and the crew of *U-30* to secrecy and ordered the boat's logbook altered. Churchill dryly noted in his memoirs that the Germans' story of his sinister hand in the sinking of the *Athenia* "received some credence in unfriendly quarters."[16]

Because of an acute shortage of escort vessels, the Admiralty originally planned to limit convoys to the east coast of Britain and to rely on other means, chiefly evasive routing and zigzagging, to protect merchant ships crossing the Atlantic "until and unless the enemy adopted unrestricted U-boat warfare," Churchill wrote. "But the sinking of the *Athenia* upset these plans."

Organizing a comprehensive system of convoys was a monumental task: on average there were 2,000 British merchant ships at sea at any given moment. Before going to bed in the early morning hours of September 5, Churchill personally wrote up the conclusions of the first Admiralty conference he had held that night, which focused on the convoy issue. The Admiralty's Trade Division, responsible for controlling shipping movements, was to set up a control room within twenty-four hours, equipped with a large chart plotting the location of every merchant vessel within three days' sailing of British shores. Destroyers were to be "scraped from the Eastern and Mediterranean theatres" to add to the available escort force. The logs of arriving merchant ships were to be examined promptly by a "competent naval authority" to make sure that ships' masters were following orders and "examples made" of serious defaulters. All independent sailings were canceled. Within a week, outbound convoys were running to the Atlantic from Liverpool and the Thames; a week later the first homeward-bound convoy sailed from Halifax, Nova Scotia.[17]

Churchill's other bold intervention in the U-boat war was a disaster. Convinced that aggressive measures were what it took to defeat an enemy, the first lord ordered the creation of "hunting groups" that would roam the oceans independently seeking out their quarry: he drew an analogy to a cavalry division. It was the wrong strategy for almost every possible reason. Although convoys had the superficial air of being passively defensive, they had a powerful offensive character in practice, forcing a U-boat to confront an escort vessel armed with asdic and depth charges if it wanted to get within striking distance of the convoyed ships. The vastness of the oceans made it an extraordinarily inefficient procedure to maraud about in a random hunt hoping to stumble on a U-boat by chance: the place to look for U-boats was where their prey was.

As a stopgap Churchill assigned two aircraft carriers, *Courageous* and *Ark Royal*, to the U-boat hunt. They were clearly the wrong tools for the job. *Courageous* was a converted First World War battle cruiser equipped with obsolescent Fairey Swordfish biplanes. The Royal Navy's aircrews had no experience hunting submarines and their torpedoes and bombs were useless for attacking a U-boat once it had dived—equipping aircraft with depth charges was an idea still in the future. But Churchill was in an improvising mood as he cast about for anything that might be rushed into the fight.

On the morning of September 18 he traveled to Scotland to inspect the Home Fleet, anchored at Loch Ewe in the Western Highlands. He wrote in his memoirs that he felt "oddly oppressed" by his memories that day. "No one had ever been over the same terrible course twice with such an interval between. No one had felt its dangers and responsibilities from the summit as I had or, to descend to a small point, understood how First Lords of the Admiralty are treated when great ships are sunk and things go wrong." Arriving the next morning at Euston Station in London on the overnight sleeper, he was surprised to see the first sea lord, Admiral Dudley Pound, on the platform waiting for him. Pound's face was grave. "I have bad news for you, First Lord," he said. "The *Courageous* was sunk yesterday evening in the Bristol Channel."[18]

Churchill would describe it as a "hundred-to-one chance" that the carrier had "happened to meet a U-boat" just at the moment she was turning into the wind and slowing to recover her aircraft, but meeting U-boats was after all what she was supposed to be doing, and sinking them; instead she was hit in quick succession by two torpedoes fired by *U-29*. An enormous secondary explosion followed the impact of the second torpedo. In twenty minutes *Courageous* went down in a calm sea, taking 518 of her crew of 1,260 to their deaths.

In his speech to Parliament a week later on the U-boat war, Churchill said, "Risks have to be run all the time in naval war, and sometime grievous forfeit is exacted." This was true enough but missed the point when the risks were yielding so little in return. He confidently asserted that "six or seven" U-boats had already been destroyed, calling that a "safe figure" and "probably an understatement." After the war it would be clear that the real figure was two. He also confidently asserted that history was not repeating itself, for all the obvious parallels:

> . . . the U-boat attack upon British ocean-wide commerce was one of the most heart-shaking hazards of the last war. It seemed during the early months of 1917 that it might compass our total ruin. Only those who lived through it at the summit know what it was like. I was at that time not in office, but . . . I watched with a fear that I never felt at any other moment in that struggle the deadly upward movement of the curve of sinkings over the arrival of new construction. That was, in my opinion, the gravest peril which we faced in all the ups and downs of that war.

"We have no reason, upon the information and experience which are now available," he concluded, "to suppose that such a situation will recur." As he would later admit, "I had accepted too readily when out of office the Admiralty view of the extent to which the submarine had been mastered." Still, in that September speech to Parliament he also spoke frankly of the "cruel and ruthless" fight that lay ahead. "Such is the U-boat war—hard, widespread

and bitter, a war of groping and drowning, a war of ambuscade and stratagem, a war of science and seamanship."[19]

ALTHOUGH THE BRITISH CONCLUSION that Germany had once again
embarked on unrestricted warfare with the sinking of the *Athenia* was premature, it was a perfectly accurate perception of German intentions. "The
German leopard does not change its spots," concluded the *Daily Record &
Mail*, and even *The Times*, its editorial writers at last permitted to pass
unpleasant remarks about the German character, suggested that this violation of the Submarine Protocol had afforded the world "yet one more
opportunity of judging exactly what value is to be placed on the most sol-
emn German promises."[20]

On September 23, 1939—just three weeks into the war—Hitler discussed
with Raeder gradually removing the restrictions placed on U-boat commanders by the laws of naval warfare. He agreed to one immediate change:
ships transmitting the *SSS* signal could now be torpedoed without warning,
though "rescue of crews is still to be attempted."

In a series of incremental steps over the next two months, all remaining
pretenses of abiding by the requirements of the Submarine Protocol and the
German navy's own official prize regulations to halt, board, and remove the
passengers, crew, and ship's papers of merchant vessels before sinking them
were abandoned. At the September 23 conference with Raeder the Führer
had made plain his intention to do so; he was only waiting to see for sure "if
the war against France and England has to be fought out to the finish." Hitler suggested various pretexts. Merchant ships definitely identified as British
or French could be fired upon without warning "since it may be assumed
they are armed." He cautioned that "the notorious expression 'unrestricted
submarine warfare' is to be avoided"; instead, the phrase "the siege of England" would be used, as a way to "free us from having to observe any
restrictions whatsoever on account of objections based on International
Law." To avert international criticism, "a neutral ship should occasionally
be treated especially well in order to show that the system has not been fundamentally altered."

It was of course complete, cynical nonsense, as Dönitz knew full well.
Two days after Hitler's meeting with Raeder, Dönitz set down in a staff
memorandum his intention to give U-boats permission to sink without
warning any ship, including neutrals, running without lights. "U-boat com-

manders would be informed by word of mouth and the sinking of a merchant ship must be justified in the war diary as due to possible confusion with a warship or auxiliary cruiser," he noted.[21] The very fact that Dönitz was ordering the war diary faked on this point shows he knew perfectly well that running without lights offered no valid legal basis for disregarding the protections accorded merchant vessels under international law. The laws of war made no exceptions for ships without lights, ships in escorted convoys, ships sending distress calls, even ships armed with defensive weaponry. As a matter of long-settled admiralty law, a merchant ship had a clearly recognized right to resist capture by an enemy cruiser—and it was solely at the warship's own risk to proceed in the face of such resistance. Merchant vessels had always had the right to carry guns for self-defense and, in time of war, to use them to try to fight off an enemy cruiser or privateer. A merchant ship that persistently refused to obey the commands of a warship to halt, or that tried to defend herself by firing her guns, certainly could be fired on in return, but even that did not permit the attacker to jettison the rules of war; upon ceasing resistance, the merchantman still had to be boarded and her crew and papers removed to a place of safety before being sunk.

The legal reasoning behind this was not mysterious, though it did turn on an anomaly in the laws of war that had always treated civilians at sea differently from those on land. Enemy civilians captured on the high seas could legitimately be held as prisoners of war, for example, and their private property taken as a lawful prize even though they were not combatants; on land, civilians and their property were to be respected. Likewise, a civilian who fired on an enemy soldier on land was a franc-tireur who could be executed as an unlawful combatant; at sea, the civilian crewmen of a merchant ship had a right of self-defense that they could exercise without fear of punishment or reprisal, and without altering their status and rights as noncombatants.[22] But the Nazis were masters at plausibly running roughshod over niceties of law.

At the end of September, coinciding with the last collapse of Polish resistance, Hitler had another long talk with the as yet indefatigable Birger Dahlerus. The Swede was soon on his way to London with yet another peace feeler from the Führer. "If the British actually want peace, they can have it in two weeks—without losing face," Hitler assured Dahlerus. Simultaneously, William Shirer recalled, "the German press and radio launched a big peace offensive" echoing the Führer's promises. After all, why should Britain and France fight any longer? Poland, "this ridiculous state . . . will never rise again," the Führer declared. He sought nothing but peace; he had no more claims to make, save for the adjustment of Germany's overseas colonies; it was, said Shirer, "the old gramophone record being replayed for the fifth or sixth time."

At a meeting of the War Cabinet to discuss Dahlerus's latest message, Halifax argued that "we should not absolutely close the door" on Hitler's latest approach. On October 12 Chamberlain did just that, with another show of the resolve that had been missing a year earlier. No reliance, he told the House, could be placed on the words "of the present German Government . . . acts—not words alone—must be forthcoming."[23] Hitler now had his answer whether Great Britain would "fight to the finish." Pausing only long enough for his propagandists to muddy the waters a bit, he proceeded to make good on his promise to Raeder to remove the remaining restrictions on the U-boat force. Throughout October and November, German statements repeated again and again the assertion that by supplying deck guns to British merchant ships and instructing them to transmit signals when attacked, the British Admiralty had made the merchant marine an arm of the Royal Navy, subject to attack without warning like any other combatant. In consequence, it was simply a matter of reality that "in waters around the British Isles and in the vicinity of the French coast the safety of neutral ships can no longer be taken for granted."[24]

Churchill warned Pound that "one must expect a violent reaction from Herr Hitler" to Britain's rejection of his "peace" initiative. "Perhaps quite soon."[25] The reaction was already in train. On the night of October 13, *U-47* slipped into Scapa Flow, the main British naval base, in the Orkney Islands off the far northern coast of the Scottish mainland. Dönitz had personally selected Kapitänleutnant Günther Prien, a hard-charging and experienced officer, thirty years old. His crew, told of their objective only once they were at sea, viewed it as a suicide mission. But so unprepared were the commanders of the British fleet for the possibility that a German submarine could penetrate their Gibraltar of the north that even after the first torpedo fired by *U-47* struck the battleship *Royal Oak*, causing a small explosion, they did not send their ships to battle stations, attributing the blast to spontaneous combustion in the ship's paint locker or some other accident aboard the ship itself. Prien reloaded his torpedo tubes and thirteen minutes later fired another salvo, this time striking his target amidships. A huge orange fireball erupted from the ammunition magazine and the 29,000-ton ship was lifted into the air. Within fifteen minutes the mighty ship rolled over and sank.

Two thirds of her 1,200-man crew were killed. *U-47* returned home safely to a hero's welcome for "the Bull of Scapa Flow" and his men.

Churchill acknowledged this "feat of arms" by the enemy. That was one of those chivalrous turns of phrase he had a fondness for, and which would soon become anachronistic in naval warfare. Seizing the moment to press the case with Hitler, Raeder on October 16 presented an "Economic Warfare Plan" that the naval staff had prepared to make the maximal case for unrestrained U-boat warfare:

> No threat by other countries, especially the United States, to come into the war—which can certainly be expected if the conflict continues for a long time—must lead to a relaxation of economic warfare once it is begun. The more ruthlessly economic warfare is waged, the earlier it will show results and the sooner the war will end.[26]

Hitler agreed. The same day he approved torpedoing any merchant ship "definitely recognized" as British or French without warning, as well as passenger ships in convoy "a short while after notice has been given of the intention to do so." A few days later the area where any darkened ships could be attacked without warning was extended to 20° west longitude, a meridian even with Iceland. On November 10 Raeder proposed lifting all remaining restrictions on torpedoing enemy passenger liners; an order to that effect was issued a few days later. And in late November or early December Dönitz issued standing order No. 154, which instructed U-boat commanders to put aside any remaining compunctions over the fate of the crews and passengers of the ships they torpedoed:

> Rescue no one take none with you. Have no care for the ships' boats. Weather conditions and the proximity of land are of no account. Care only for your own boat and strive to achieve the next success as soon as possible! We must be hard in this war. The enemy started this war in order to destroy us, therefore nothing else matters. . . . There are situations in attack where one could have grounds for giving up. These moments or feelings must be overcome.[27]

As British reaction to the *Athenia* sinking showed, there was little surprise in the discovery that Hitler's Germany would be no less brutal than the Kaiser's; the only wonder was that anyone could have thought otherwise. The Nazi Party's nihilistic glorification of brutality as part of its cult of power had been evident from its very beginnings, with storm trooper brawls on the streets of Munich. The only thing that had changed since the First World War in Germany's plans for unrestricted submarine warfare was Hitler's greater skill in manipulating world public opinion.

CHURCHILL TURNED the elegant library on the top floor of Admiralty House into his war room, and so that he could be at his post at all times he had the attics above converted into a flat, which he moved into at once. Installed in an office next door to the war room was Lindemann, the first lord's "tame scientist," as one fellow cabinet member sarcastically termed him.[28]

The scientific folie à deux of Churchill and Lindemann if anything grew worse. Churchill now had a whole ministry at his beck and call to carry out his enthusiasms, which became a source of despair to those who had to obey his whims. "Anything unusual or odd or dramatic intrigued him . . . deception, sabotage, and no doubt influenced by Professor Lindemann, the application of novel scientific methods," lamented Rear Admiral J. H. Godfrey, the director of naval intelligence and one of the more capable and scientific-minded men in the navy's senior ranks. Churchill immediately ordered full-scale development of the aerial mine-curtain scheme that the Tizard Committee had repeatedly vetoed as scientifically impractical. The device had now evolved into a contraption with multiple rocket launchers that would send a projectile up to 2,000 feet and release a curtain of long wires, each with a parachute at the top end and a bomblet at the bottom. The idea was that an attacking enemy plane would snare one of the wires with its wing; the drag from the parachute would then yank the wire back until the bomb contacted the front edge of the wing, when it would detonate. Churchill dubbed it the "UP Weapon." UP stood for "unrotated projectile," a name intended to fool the Germans as to its true brilliant nature. He ordered forty of the devices built and had them installed on warships even though, as one admiral bluntly observed, the gadget "was considered by everyone except Winston as plain crazy." It was in fact worthless, less than worthless even: the rockets were slow to fire, the wires were plainly visible in the air and easily avoided by enemy aircraft, and the unexploded mines even endangered the ships that fired them when they drifted back down. The launchers took up deck space that would have been far better given to guns or other weapons that worked.[29]

Churchill tried to galvanize research to counter the U-boats with more of his amateur inventions. "I wondered whether it would not be possible, when U-boats are suspected of being sunk in fifty or sixty fathoms," he wrote the vice chief of the naval staff in February 1940, "to let down a cable with a magnet attached, which magnet would be attracted to the steel mass lying on the bottom and would prove that something lay there. I asked Professor Lindemann for a note on this. Perhaps one could have an apparatus which fired a charge or made a signal."[30]

Churchill also had Lindemann, "my old friend and confidant of so many years" he called him in his memoirs, create a sort of personal statistics department within the Admiralty to provide him "my own sure, steady source of information." The need was undeniable: there was no overall government statistical organization and Churchill rightly understood and appreciated the importance of getting the numbers and having them presented in clear tables and graphs. But even here Churchill's overreliance on a single trusted protégé as the arbiter of all scientific and mathematical wisdom at times misfired. In preparing statistics on the loss rates of ships traveling in convoys versus those traveling independently, Lindemann arbitrarily ruled that ships which had been in convoy but were sunk after they dispersed and traveled independently on the final leg of the outbound voyage should still be counted as a loss-in-convoy. "This statistical error prevented for a long time a true comparison," noted Captain B. B. Schofield, who took over the Admiralty's Trade Division in March 1941; it encouraged advocates of independent sailings in their mistaken belief that the costs of convoying (in delaying the movement of goods) did not outweigh the benefits (in lower loss rates). As Schofield would note, a more complete statistical analysis subsequently left no doubt that only the very few extremely fast cargo liners capable of speeds greater than 20 knots were better off going it alone across the Atlantic.[31]

But even those driven to distraction by Churchill's brainstorms tended to view them as a sort of transaction cost; they were just the price one had to pay for Churchill's "demonic energy and extraordinary imagination," as Godfrey put it. Occasionally they hit the jackpot. A Churchill-Lindemann scheme for defeating magnetic mines by "degaussing" the metal hulls of warships, wrapping them with coils of electrical wire to form a giant electromagnet that countered the magnetism of their steel, was a brilliant success. Churchill's diversion of scarce warships to "hunter groups" was a grievous blunder, but he also fully understood the urgent need for convoy escorts. In a few swift strokes Churchill ordered construction to begin at once of small escort vessels that could quickly make up the shortfall. He revived the old names *corvettes* and *frigates*, with their dashing associations of the age of sail, to designate these new types, which were about two thirds the size and displacement of a regular destroyer. The first, he hoped, could be ready in twelve or even eight months.[32]

Churchill had also fully grasped the revolutionary import of radar when much of the Admiralty had not. Here was an instance where his belief that navies always had to be dragged into the modern world against their will had more than a little merit. Tizard's committee had tried to bring the Admiralty in on radar from the start, and had met almost comical opposition. Not long after the first radar experiments in February 1935 Tizard realized that however important radar was to the air force it must be as important or even more so for the navy. "Full of this," he recalled, "I sought an early interview with a naval expert." He told him about the work, explained its significance, pointed out that radar could in principle make it possible to locate and accurately direct gunfire on an enemy ship even beyond visible range, a development with truly revolutionary implications for naval warfare. "He listened to me patiently," Tizard remembered, "then he said, 'May I ask if you have ever seen a warship?' I said, 'Yes, many times.' 'And have you observed the aerials?' 'Yes,' I answered. 'Well,' he said, 'if you had observed them closely enough you would know that there was no room on them for any more.'" Tizard replied that perhaps some of them might be removed.[33]

Little substantive progress had been made since. The navy insisted that it needed its own radar research laboratory to address the special problems of operating and maintaining the equipment at sea, and had engaged in a lengthy bureaucratic fight with Watson-Watt, who wanted all radar research centralized at the air force facility he directed at Bawdsey. The navy won, but funding remained pathetic and by the outbreak of war only two operational warships, the cruiser *Sheffield* and the battleship *Rodney,* were equipped with radar (known at the time in Britain by the cover name "radio direction finding," RDF for short). Those first shipborne units were designed to scan the skies for enemy aircraft. But Churchill was well aware that radar also offered the only hope of detecting surfaced submarines at night. In his first week on the job at the start of September 1939, he confronted Rear Admiral

Bruce Austin Fraser, the third sea lord and controller of the navy, responsible for naval construction, about the matter. "The fitting of R.D.F. in HM
ships, especially those engaged in the U-boat fighting, is of high urgency,"
Churchill wrote. "Do you know about this? If not, see Professor Tizard,
and put him in contact with Admiral Somerville."[34] The first anti–surface
vessel, or ASV, radars would begin arriving in spring 1941. Installed on the
corvettes and other escort vessels, they made an immediate impact on the
battle—especially given the tactical importance that night surfaced attacks
assumed in Dönitz's plans.

Churchill also told Fraser, in a September 28 minute headed "Urgent," to
assist the RAF with some experiments it wanted to begin to see if airplanes
could carry and drop depth charges. The idea apparently had not occurred
to anyone in the navy, but Churchill had no time for turf battles: "Will you
kindly make available for Commander Anderson RAF sometime tomorrow
a depth-charge case, empty, together with a statement of the weight with
which it should be filled, so that the experiments can be made. . . . Commander Anderson will call for it, and take it to his Squadron."[35]

The real problems with bringing science into the war effort went far
deeper than anything that one energetic first lord, and his tame scientist,
could possibly accomplish by themselves even on their good days. The
Admiralty, and the anti-U-boat effort in particular, simply lacked either the
organization or the institutional culture to make use of scientific expertise
in any systematic fashion. The radar program was about the only thing on
the scientific front that *was* moving relatively quickly. In the first week of the
war the Admiralty Signal Establishment enrolled 200 physicists and electrical engineers from the Central Register list, bypassing the usual civil service
procedures and arranging for security clearances in as little as twenty-four
hours. But with that sole exception, the movement of scientists into the military bureaucracies even with the start of the war was "a slow and haphazard
business," noted Tizard's biographer Ronald Clark.[36]

Some of the fault lay in simple chaos; much lay in the fact that the conventional lines of military authority made no provision for civilian scientists to do anything more than offer advice, even about the direction and
organization of scientific research within the services. The Tizard Committee had relied almost entirely on unofficial string pulling to get its job done.
Lacking the power to order anyone in the Air Ministry to do anything, Tizard had shepherded the radar program over the previous four years with a
combination of personal political connections, friendships with top RAF

officers, and scientific contacts he, Blackett, and Hill had with colleagues in the air force's various research establishments. But such informal arrangements were completely unequal to mobilizing an army of scientific workers to fight a war.

Meanwhile, the old prejudices against scientists, not to mention intellectuals in general, ran deep through the services; as did the more fundamental impatience toward civilians who presumed to tell military men how to fight their war for them. Most officers still saw no good reason why they should hire scientists, or listen to them once they had been hired. The lingering class snobbery that equated the sciences or the possession of professional expertise with "trade" even played a part.

That was particularly evident in the British intelligence services, which had a deeply ingrained culture of genteel amateurism. The Government Code and Cypher School, responsible for breaking enemy code systems, was a sort of old boy network of its own, dominated by Cambridge and Oxford art historians, professors of medieval German, lecturers in ancient Greek, and other distinctly nonscientific types. The first mathematician to be hired by GC&CS, Peter Twinn, joined the staff of the code-breaking unit only in 1938; he was told later that there had been grave doubts about hiring a mathematician at all, as they were regarded as "strange fellows notoriously impractical." If having someone with scientific training "were regretfully to be accepted as an unavoidable necessity," as Twinn humorously described the prevailing attitude, the general thought was "it might not be better to look for a physicist on the grounds that they might be expected to have at least some appreciation of the real world."[37] Of the twenty-one academics brought in to GC&CS during the first few weeks of the war, only three were mathematicians. All of the rest were from the humanities, and came through the usual channels; they had, in the words of GC&CS's head, Alastair Denniston, been recruited by "men now in senior positions" at Oxford and Cambridge who "had worked in our ranks during 1914–1918."[38] A few things had changed in the rest of the world since then. Not least was the advent of machine-generated ciphers like the Enigma, which by the time the job was done would demand the services of literally thousands of engineers, scientists, and mathematicians to crack.

With the coming of war, GC&CS had established itself at another of those bizarre nineteenth-century manor houses built in a cacophony of architectural styles, Bletchley Park, located about fifty miles northwest of London. A tiny group under a brilliant, cantankerous, and very old-school cryptographer named Alfred Dillwyn Knox, known as "Dilly," had made some small progress on the Enigma problem before the war, but had been largely stymied. Their first major break came from an eleventh-hour hand-off of a treasure trove of discoveries made by three young code breakers—all mathematicians—in the Polish army's cipher bureau, the Biuro Szyfrow, who just weeks before the Nazi invasion summoned their British counter-parts to Warsaw to reveal what they had managed to do. Among other prod-igies, they had pulled off the mind-boggling analytic feat of reconstructing, sight unseen, the internal wiring of the machine's scrambler rotors. The mathematics required was something called permutation theory, and it was not for the mathematically faint of heart. Using these methods the Poles had been able to read German army Enigma messages on a regular basis for several extended periods going back to 1933. They had even constructed a primitive mechanical computer to help unlock each day's changing setting of the Enigma rotors and plugs.

The Poles had made some theoretical progress on the much more dif-ficult naval Enigma, but had essentially given it up in the face of the extreme difficulties that arose from the navy's use of printed code lists to choose and encode the "indicator" groups at the start of each message that speci-fied the rotor setting. At Bletchley, even after the Poles had turned over the fruits of their research, the naval Enigma problem was viewed as simply impossible—literally not worth wasting any time or effort on. Alan Tur-ing, one of the three mathematicians who arrived at Bletchley in September 1939, and who would later achieve renown for laying the mathematical and logical groundwork of the modern digital computer, began working on the naval Enigma, he admitted later, only "because no one else was doing any-thing about it and I could have it to myself."[39]

Even when the British code breakers did produce results tackling other enemy code systems during the first year of the war, they had a hard time getting anyone in the military to pay attention. The Admiralty was in many ways the worst. Harry Hinsley, a young Cambridge student, arrived at Bletchley in October 1939. By December he was, in his definitely ironic words, "the leading expert outside Germany on the wireless organization of the German Navy." It was a claim, he quickly added, that did "not amount to much," given how little work was being done. Every once in a while Hins-ley would crank the handle on a telephone that linked him to the Admiralty and report to an ostentatiously bored naval officer on the other end of the wire something he had discovered. By the spring, the Bletchley code breakers had managed to read some paper-and-pencil ciphers the German navy used for its more routine communications. In the first week of April 1940 Christopher Morris, a fellow of King's College who had joined the German navy group, deciphered a message in the German merchant navy code. It ordered ships heading for Bergen to report their positions at stated intervals to the War Office in Berlin. Morris dutifully reported this intriguing finding, and was told he obviously did not know what he was talking about, as ships report to *naval* headquarters, not army headquarters.[40]

The subsequent explanation was simple and appalling. They were troopships, and the Admiralty had just missed receiving the only warning it would have that Hitler was about to invade Norway.

ANYONE WHO LIVES THROUGH a war knows that its moments of intense excitement are no match for the endless stretches of monumental boredom. The boredom of the opening months of the war in Britain, however, seemed to be of a different class altogether: vast, encompassing, incomprehensible. The American war correspondents who had flocked to Europe to cover all the excitement promptly dubbed it the "phony war," or the "Bore War." France, her entire military plans and strategy built upon the defensive might of the Maginot Line, did nothing and waited. The Luftwaffe bombers expected to rain poison gas upon London in the opening moments failed to appear, night after night. One in five Britons, a Gallup poll reported a few months into the war, had been injured on the blacked-out streets at night, stumbling on sidewalks, hit by cars, colliding with other pedestrians. A stifling heat wave that hit London in September gave way to a cold gray fall and then to the bitterest winter in forty-five years. Coal was in short supply, water pipes burst, trains stalled in snowdrifts, the Thames froze to solid ice. Leonard Woolf compared the feeling of living in wartime Britain to "endlessly waiting in a dirty, grey railway station waiting-room, a cosmic railway station waiting-room, with nothing to do but wait endlessly for the next catastrophe."[41]

The one front where things were happening through the winter and spring of 1940 offered up regular catastrophes, but they remained hidden from most Britons by the mists and waves of the dark seas where Churchill's war of "groping and drowning" took its steady toll. Despite an epidemic of torpedo malfunctions that caused the weapons to explode prematurely or not at all, the U-boats had already taken a voracious bite out of Britain's ocean lifeline. The torpedo failures were traced to defects in the design of the magnetic pistol that was supposed to trigger the warhead when it passed under the hull of its target. It turned out that it had been tested a total of two times before the war, and that variations in the earth's magnetic field plus the rigors of wartime conditions (including diving far deeper than the U-boats had been permitted to do during training) upset the trigger's delicate mechanism. An investigation demanded by Dönitz also found that the torpedoes tended to run too deep, a fact it turned out the designers were aware of but thought would not matter given the magnetic triggering system. "The result is staggering," Dönitz wrote in his war diary. "I do not believe that ever in the history of war men have been sent against the enemy with such useless weapons." He estimated that nearly half of the unsuccessful torpedo shots had been due to failures in the magnetic pistols. "In practice the boats are unarmed," he wrote in another moment of exasperation. "Of 22 shots fired in the last few days at least 9 have been premature detonations which have in turn caused other torpedoes fired at the same time to explode prematurely or miss."[42]

Even with the failures, which Dönitz calculated had robbed him of a third of a million tons of British shipping, his U-boats had sunk nearly a million tons by the time they were withdrawn for the Norway invasion. That was with an average of only six of the oceangoing U-boats at sea each day for the first six months of the war. Being forced to institute convoys had alone resulted in a reduction of 25 percent in British imports. Even with the loss of seventeen U-boats to a variety of causes during that time, the picture for the future was distinctly encouraging to BdU. At the steeply accelerating rates of new construction currently planned, there would be ten times as many U-boats available to bring to the war against British trade by the spring of 1942.[43]

6

Blackett's Circus

IN MAY 1940 Hitler sent his panzers and 2 million men streaming into the Low Countries and France; Chamberlain resigned and Churchill became prime minister. A few weeks later the first night attacks by Luftwaffe bombers began over Britain. Meanwhile, the scientists fretted.

Shortly after the start of the war Blackett had been appointed a scientific officer at the Royal Aircraft Establishment at Farnborough, where he was now spending two days a week working to develop a new bombsight. But most of his colleagues were growing increasingly impatient at having heard nothing about the promised mobilization of scientists. "From the time of the Munich crisis in 1938," Solly Zuckerman recalled, "there had been all manner of talk . . . that given a war, all we would have to do was wait until told what our battle-stations were. Nothing happened."[1]

Zuckerman, back in 1931, had started a London dining club of like-minded scientific colleagues. It was called the Tots and Quots, derived from the Latin tag "*Quot homines, tot sententiae*": "As many opinions as there are men." The group had petered out after a few years, but with the coming of the war Zuckerman revived it; they met once a month at a Soho restaurant and usually invited a distinguished guest who would open the after-dinner discussion. Increasingly the main topic of discussion now was the underutilization of scientists in the war effort.

Zuckerman was from a South African Jewish family and had come to London in 1925 on a scholarship to complete his medical education. He almost immediately made a splash with a series of anatomical and hormonal studies of the estrous cycle of baboons in the London Zoo, and after qualifying as a doctor shifted his career completely to primatology. In 1932, at the age of twenty-eight, he published *The Social Life of Monkeys and Apes*, a standard work that would be repeatedly reprinted over the next half century. Two years later he joined the Department of Human Anatomy at Cambridge.

As a boy he had been aloof, withdrawn, and reserved; in London and Cambridge, as a rising scientific star, he seemed almost to explode with enthusiasm for his subject and for the social milieu in which he found himself suddenly immersed. From that point on he assiduously collected famous friends, from literary and artistic circles as well as scientific ones—E. E. Cummings, Alfred Hitchcock, George and Ira Gershwin, and H. G. Wells among them. A physiology student of his during this time recorded his impression of Zuckerman as "an almost incoherently enthusiastic young man who danced around and told us doubtful stories about the sexual cycles and activities of baboons. None of us guessed at the time what a breakthrough it was." Many colleagues noted how Zuckerman was, in the words of one, "dazzlingly quick to grasp a point"; in fact he usually could grasp a point faster than he could explain it. Another said: "He was intolerant of people he regarded as less clever than himself—a very large group."[2]

The Tots and Quots had an unmistakably young, left, and bohemian slant. Blackett was a regular member, as was Bernal, and the evolutionary biologist Julian Huxley. Another regular was Conrad Hal Waddington, a biologist and fellow of Christ's College, Cambridge. Waddington was another polymath. Brought up in England by relatives from age four while his parents remained on their tea estate in India, he devoted himself to a series of enthusiastic pursuits including hunting for fossils and conducting chemistry experiments as a boy; at university his extremely serious interests included Morris dancing, rock climbing, avant-garde art and architecture (his friends included Henry Moore, Alexander Calder, and Walter Gropius), and philosophy (he would recount long and mysterious debates he had at Cambridge with Ludwig Wittgenstein about the nature of language and reality). In his later career, after the war, Waddington would become a leader in the field of evolutionary genetics, making major experimental and theoretical contributions, though as a young scientist he was by his own reckoning somewhat adrift and unfocused. He had become almost completely bald at age twenty-one, which had the effect of making "many people think he was much older than he actually was," recalled one friend. "Sometimes, when throwing out half-formed hypotheses or ideas he seemed surprised at being taken seriously."[3]

For the June 12, 1940, meeting of the Tots and Quots, Zuckerman invited as a guest Allen Lane, the publisher of Penguin paperbacks. It would prove to be one of those quirky pivotal events that change everything. It certainly changed Zuckerman's future. "I was moved along by one accident after another with little idea of who I was or of what I would become and with little notion of what the morrow would bring," Zuckerman later said. "Up to the time of the Second World War I should have laughed if anyone had suggested that in the years ahead I would become involved in public events."[4] Lane was fascinated by the discussion that took place about what science could do to help win the war, and on the spur of the moment made the group an offer. If they would write up their argument—and get it to him in two weeks—he would publish it immediately. The scientists delivered their manuscript eleven days later and in late July a small 140-page "Penguin Special" appeared bearing the title *Science in War*. On a plain orange and white cover was a block of stark text that began: "The full use of our scientific resources is essential if we are to win the war. To-day they are being half used."

As Julian Huxley subsequently noted in a review in *Nature*, the book's one-month production schedule was an "abbreviated gestation more characteristic of a rodent than of a human being or a book."[5] But Lane's instincts had been right: the timing could not have been better. *Science in War* reached bookstores and newsstands just as the German air attack on Britain was beginning in earnest. Priced at six pennies, it immediately sold thousands of copies. The book largely restated the case the scientific left had been making for years. Industry and society had failed to put to use the advances of science that could vastly increase production and efficiency. The war now made it more urgent than ever—essential, in fact—for scientific methods to be brought to bear on every aspect of society. Indeed, the strains created by wartime shortages of materials and skilled labor and the disruption of normal business made it all the more impossible to leave decision making to tradition, gut feeling, emotion, or guesswork.

The authors gave numerous examples of how science could be put to work: improving public and occupational health to increase workers' productivity; using scientific organizational concepts to streamline management and optimize factory workflows; providing better treatments for wounded soldiers; developing new methods of mass food production such as aquaculture; boosting meat and milk yields with growth hormones. Some of the ideas veered, slightly chillingly, into Orwellian social engineering. Psychological methods could be put to use to produce more effective home front propaganda; communally organized kitchens and feeding stations could eliminate the wasteful duplication of home meal preparation. But mainly the authors emphasized that effective use of science, especially in wartime, could come about only through vigorous government planning and direction. The entrenched attitude of "Victorian Liberalism," bureaucratic caution, and "*laissez-faire* . . . Government non-interference" that characterized the civil service was the major obstacle that had to be overcome.[6] It was, in short, a call for government to take charge and put science to work in the interests of all, just as Blackett had urged in his 1934 BBC broadcast *The Frustration of Science*. The war now provided the justification that socialist idealism had, as yet, failed to.

SCIENCE IN WAR was published anonymously, the authors described only as "twenty-five scientists." Blackett almost certainly wrote the section of the book devoted to operational research. The accomplishments of Eric C. Williams's operational research group at Bawdsey had strongly reaffirmed Blackett's conviction that this was a model for a greatly expanded role for scientists in the war. During the 1939 summer air exercises, which were the first complete test of the radar defense system in the hands of RAF operators, Williams's team had closely observed the work of the "Filter Rooms," where data from the various radar stations was correlated. The Filter Rooms had the job of reconciling and triangulating sometimes contradictory radar data to determine the height, location, direction, and size of an incoming raid. The scientists on the scene were able to spot bottlenecks, devise procedural changes, and introduce some simple probabilistic rules to help the operators eliminate data that was likely to be erroneous.

A second scientific team was stationed at the fighter group operations rooms where the tracks passed on by the Filter Rooms were plotted and orders issued to the fighter squadrons under their command; the members of this team, too, found themselves focusing as much on organization and procedures as on technical matters directly related to the operation of the radar equipment. One key problem they pinpointed was that tracks were frequently lost when a raid passed from one section to another, and they recommended giving one control room officer the sole job of maintaining continuity.

The radar researchers at Bawdsey were evacuated to Dundee in Scotland at the start of the war but the operational research teams had already sufficiently impressed the top officers at Fighter Command with their on-the-scene indispensability that Air Chief Marshal Dowding, the commander-in-chief, asked for them to be kept behind. The two groups were combined in a single Operational Research Section (as it would later be renamed, in 1941) under the direction of a Canadian physicist, Harold Larnder, at Stanmore, the Fighter Command headquarters located at one of the highest points of London, about ten miles northwest of the city center. Dowding gave the ORS his unreserved support, sending Larnder appreciative notes every time one of the sporadic Luftwaffe daylight raids that took place through the fall and winter of 1939–1940 was successfully intercepted—almost as if the ORS had accomplished the feat itself. A postwar RAF report concluded, "The high state of efficiency reached by the radar stations by the time of the Battle of Britain was to a large extent due to the fact that, from the time of the Firth of Forth raid in October 1939 onwards, the ORS analysed almost every failure to intercept daylight raids," and figured out ways to improve performance the next time.[7]

In January 1940 Tizard suggested to Rowe that an expanded OR section might be created to study questions for the air force not immediately connected with the radar system. "I had hoped that Blackett would be able to devote a lot of time to the work," Tizard wrote, "he is the ideal man for the job."[8] Nothing immediately came of the suggestion, but the idea of applying science to war at a very fundamental level clearly was much on Blackett's mind at the time. Zuckerman would later claim with only slight exaggeration that it was the Tots and Quots—and *Science in War*—that brought operational research to the fore: "Operational research was, to a significant extent, the creation of our members."[9] The passage in *Science in War* making the case for giving to scientists a comprehensive role in the development of military strategy and tactics did not use the term "operational research," but that was what it was unmistakably referring to:

In the actual business of warfare, science has been used up to now almost exclusively on the technical side—for example, to improve weapons, transport, and communication. It has hardly been used, at least by us, on the more general, and the more vitally important question of strategy and tactics. These, on paper, depend on the special discipline of military science, which, however, has little or no relation to the natural and social sciences. The true scientific departments of the Services concentrate either on detailed technical problems as they arise, or on general technical questions, such as the improvement of the ballistic properties of guns, or of the speed and fighting power of aeroplanes. Yet the use of these weapons and the organization of the men who handle them are at least as much scientific problems as is their production. The waging of warfare represents a series of human operations carried out for more or less definite ends. Seeing whether these operations actually yield the results expected from them should be a matter of direct scientific analysis. The ultimate answer is provided by victory or defeat, but failure to understand the factors contributing to that victory or defeat, and the degree to which each contributes, removes any secure ground for organizing further success. A naïve belief in invincibility may have some value in morale, but, as experience in France has shown, it is a dangerous guide in strategy.

It is possible to reduce many of the factors in military operations to numerical values. Doing so provides problems capable of definite solution. This has, indeed, been done to a certain extent with the tactical problems of naval and air fighting, but it could be extended to many more. The scientific staffs of the Services need to play a much larger part than they seem to do in the formulation and solution of strategical and tactical problems. . . . The disadvantage until now has been that scientists of sufficient ability have not been made aware enough of the situation to be of any greater use than other amateur strategists. There is, however, little doubt that geographic and economic knowledge, and the assistance of great modern developments in mathematics, could lead, in a minimum of time, to a revolution in strategy far greater than that introduced by Napoleon.[10]

The anonymity of the book's authors allowed Zuckerman, Huxley, and J. G. Crowther, a member of the Tots and Quots who was the science correspondent for the *Manchester Guardian*, to shamelessly plug their own book in laudatory reviews they published in *Nature* and the *Guardian*. Whatever the ethics of that, it helped to secure attention in high places. Within weeks of the publication of *Science in War*, urgent questions were being raised in Parliament about the use of scientists in the war.[11]

———

TWO DAYS AFTER the surrender of France on June 21, 1940, a convoy of trucks departed the German U-boat base at Wilhelmshaven. Loaded with torpedoes, mines, spare parts, and supplies, its first stop was Paris, where Dönitz had temporarily set up his headquarters while he personally conducted a survey of the French ports on the Bay of Biscay. A few days later the convoy headed on to L'Orient. As a base for the U-boats, the port was 450 miles closer to the Atlantic shipping lanes. Dönitz commandeered for his new operating headquarters a splendid villa near one of the eighteenth-century stone forts that commanded the north bank of the river opposite L'Orient. Lemp's *U-30*, the boat that had torpedoed the *Athenia*, was the first to put in at the new base, on July 7. Other Biscay ports were soon readied to receive the U-boats: Brest, Bordeaux, Saint-Nazaire, La Pallice.

With the Royal Navy's destroyers withdrawn to the Channel, standing vigil against Hitler's threatened invasion of England, the Atlantic was left wide open to German submarines. Dönitz had ordered the faulty magnetic torpedo pistols replaced with a redesigned contact detonator, and on a five-week cruise in late May and early June *U-37* successfully dispatched four ships with the revamped torpedoes, restoring confidence to U-boat commanders throughout the fleet.[12] Meanwhile the Kriegsmarine's radio intelligence bureau, the Beobachtungsdienst ("observation service," B-dienst for short), was supplying a steady stream of information on British convoy sailings and routes. Due to appallingly sloppy cipher-security procedures employed by the Royal Navy, the B-dienst had long been able to read all but the most high-level British naval radio traffic. British communications with convoys and escorts employed a relatively simple code book system known as the Naval Cypher, and the B-dienst was routinely breaking 50 percent of these messages during this period.[13] Dönitz had further honed his command procedures to emphasize central control of all U-boats directly by BdU headquarters in order to organize a "wolf pack" attack once a convoy was located. The confluence of all of these forces in the summer of 1940 swept a tide of good fortune to the U-boat service. In the three months from July to September the U-boats sank more than 150 ships, close to a million tons in all. The U-boat commanders had a name for this period: the "Happy Time."

Just as in the previous war, the U-boat captains presented to the German public the sort of heroes otherwise missing in a war of anonymous mass mobilization. Not that there was anything terribly chivalrous about firing torpedoes in the dark at unsuspecting civilian cargo ships; nonetheless, the U-boat commanders were young and daring and individually identifiable, which was what mattered. Goebbels's propaganda celebrated them as the "gray wolves." "Aces" who sank 50,000 tons were showered with medals, and more publicity. Günther Prien of Scapa Flow fame was among the first, sinking eight ships in one two-week period in late June. Otto Kretschmer of *U-99*, sailing from L'Orient at the end of July, a golden horseshoe riveted to his conning tower for good luck, set a record by sinking seven ships totaling 65,000 tons in a single two-week period.

Though many later writers (and many of the surviving U-boat officers themselves, after the war) would portray the U-boat men as nonpolitical professional soldiers who even questioned Hitler and the Nazi leadership, they were, at least at the start, as devoted and fanatical as they come. In the beginning the U-boat crews were still all volunteers, their ardent pro-Nazism mirroring their personal swagger and arrogance. The British interrogators of one captured crew noted, "The usual undigested propaganda was repeated *verbatim* and *ad nauseam*." The hero treatment in turn cemented their loyalty to the regime. When Kretschmer's *U-99* was sunk the following March of 1941, the British interrogation report of the captured Germans observed: "The crew . . . had an exaggerated idea of their importance and dignity; these inflated opinions were no doubt due to the extraordinary degree of public adulation to which they had become accustomed." (The interrogators did, however, find that Kretschmer's own political views "were less extremely Nazi than had been assumed," and though a national hero he had "become weary of the war some time ago." The junior officers were generally more "typical Nazis.")[14]

With their extra allowances, the U-boat crews earned about twice the pay of other officers and men in the German navy, and Dönitz saw to it that ashore they received the best provisions, rest and relaxation in requisitioned luxury hotels and chateaux in the French countryside, and the chance to blow off copious quantities of steam. Raucous parties, rivers of champagne, and accommodating Frenchwomen were always available in abundance. In L'Orient a favorite haunt was a bar called Les Trois Soeurs, known to the U-boat crews as Die Sechs Titten. "Living like a God in France" was how the submariners described their life between cruises.[15] During cruises, they had an excellent opportunity of dying for the Führer: one in three U-boats did not even make it back from their first war patrol.

WITH CHURCHILL'S MOVE to No. 10 Downing Street, Lindemann's influence reached new—and to Tizard and many of the other scientists maddening—heights. Tizard was now officially the principal scientific adviser to the Air Staff as well as chairman of the Committee for the Scientific Study of Air Warfare, but Lindemann started showing up at his meetings and stepping on his toes constantly. Tizard, Blackett, and A. V. Hill rightly saw that Lindemann meant to refight the match he had lost against them in 1936 when he had been ousted from the air defense committee; now, with the prime minister's backing, Lindemann clearly was out to wrest control of all government scientific advice from his old rivals. In particular it was clear he meant either to cut out the CSSAW's influence in scientific decisions or simply take effective control of the committee himself. He continued to push his aerial mine scheme and other slightly crackpot ideas, notably a plan to employ incendiary phosphorous pellets to set fire to German crops—a scheme that the War Cabinet, without any serious discussion, approved £1 million to develop. Tizard believed that on technical grounds alone it would almost certainly fail, never mind the moral questions it raised. But with the prime minister's ear and confidence, the Prof's dogmatic and opinionated views increasingly had no check.

Hill drew up a secret memorandum which was endorsed by Tizard and other leading scientists alarmed at the Rasputin-like role Lindemann had secured for himself at Churchill's side:

> It is unfortunate that Professor Lindemann, whose advice appears to be taken by the Cabinet in such matters, is completely out of touch with his scientific colleagues. He does not consult with them, he refuses to co-operate or to discuss matters with them, and it is the considered opinion, based on long experience, of a number of the most responsible and experienced among them that his judgment is too often unsound. They feel indeed that his methods and his influence are dangerous. . . .
>
> I realise that Professor Lindemann's presence may be indispensable to the Prime Minister, and the prestige and influence of the Prime Minister are now so important to the nation that some compromise may be necessary. It is impossible, however, for the present situation to continue. . . . The whole trouble is not due to one person, but that a system has grown up of

taking sudden technical decisions of high importance without, or against, technical advice.[16]

Disastrously, just as this confrontation with Lindemann was coming to a head in the summer of 1940, Tizard made a rare scientific blunder. Worse, he did so in Churchill's presence, and it largely sealed his fate. The issue was the growing evidence, though still highly speculative, that the Germans were using some sort of highly accurate radio guidance device to steer their bombers to their targets in night raids over British cities. Lindemann, with his free-roving brief from the prime minister, had summoned R. V. Jones, his former student who was now looking into the problem at the Air Ministry, and picked his brains. Lindemann sent a note to Churchill outlining the evidence. "With your approval I will take this up with the Air Ministry and try to stimulate action," Lindemann wrote. Churchill scrawled a reply at the bottom: "This seems most intriguing and I hope you will have it thoroughly examined."[17]

Jones had had his own run-ins with Lindemann while working at his Clarendon Laboratory at Oxford, and had experienced firsthand Lindemann's stubborn insistence on sticking to erroneous conclusions once he had leapt to them and his arrogant dismissal of technical objections to his ideas as "defeatism." He also was deeply indebted to Tizard, having been hired to head the government's scientific intelligence program as a direct result of the Tizard Committee's recommendation: Noting with concern the near-complete absence of any information about the Germans' efforts to apply science to military use, the committee in April 1939 had convinced the Air Ministry to create a new staff position dedicated to that task, and proposed Jones for it.

Jones, though only twenty-seven at the time, was in many ways the perfect candidate for the job. He had a solid grounding in experimental physics. He also had a brash and outgoing personality, along with an extremely quick sense of humor that he liked to put to the test with high-risk practical jokes. One day, while discussing the many possibilities of telephone hoaxes with a fellow physics graduate student at Oxford, he offhandedly suggested it ought to be possible to induce the victim of such a joke to place his telephone into a bucket of water. His friend immediately proposed to take him up on the challenge. They agreed that Jones would ring the friend's flat in a few minutes; the friend rushed home to observe the proceedings. After first ringing and hanging up a few times to simulate some trouble with the connection, Jones got one of his friend's flatmates on the line. He explained in a convincing Cockney accent that he was a telephone engineer; a fault had been reported on the line; they would not be able to send someone out until the next day to repair it; it was just possible, however, that they could fix the trouble by performing a few tests right now. Would he have a few minutes to help?

The victim obligingly agreed, and Jones proceeded to lead him through a series of increasingly absurd "tests," supposedly to trace a leak to ground that he said seemed to be shorting out the line. Finally Jones said that one final test was required; for this they needed a good ground. "Have you got such a thing as a bucket of water?" he asked. The victim was just about to place the phone in the bucket as directed by Jones when his flatmate decided things had gone far enough and tried to pull the receiver out of his hand. A struggle ensued until he finally shouted, "It's Jones, you fool!" (The victim—a "manifest sportsman," in Jones's words—collapsed in laughter.)[18]

The point was that Jones was not only an able physicist; he also was quick on his feet and instinctively grasped human nature and psychology. When offered the scientific intelligence job he had leapt at it: "A man in that position could lose the war," he exclaimed. "I'll take it!"[19]

The evidence for the German's radio guidance system was far from certain. But Jones brilliantly pieced together a case from a string of circumstantial clues. A scrap of paper found in a Heinkel 111 bomber that had crashed in England in March 1940 contained a cryptic reference to a radio beacon called "Knickebein." A decoded German radio signal also mentioned Knickebein along with a particular latitude and longitude corresponding to a point over England. Two German prisoners taken from another crashed He-111 were interrogated about Knickebein and revealed nothing, but when they were alone afterward they were secretly monitored, and one was overheard saying that no matter how hard the British searched the plane "they would never find it."

That last clue had prompted Jones to ask the engineer who had gone over the German bomber if there was anything odd about the plane's Lorenz receiver, a standard device used for "blind landings" at night and in bad weather. The Lorenz system worked by broadcasting a narrowly focused radio beacon from a transmitter at the end of an airport runway; a receiver in an approaching airplane detected the signal and gave the pilot an indication if the plane was centered along the line of the radio beam, and thus in line with the runway, or to the left or right of its proper course. Jones's notion was that the same basic concept could be used to guide the German bombers as they flew *away* from the beacon, following the radio beam toward their targets in England. The Lorenz receiver was the only piece of otherwise innocent-looking equipment he could think of that might conceivably be suitable for such a "blind bombing" guidance system. The engineer pondered Jones's question for a moment and replied, "No . . . But now that you mention it, it *is* much more sensitive than they would ever need for blind landing."[20]

Tizard had largely been cut out of the discussion. When he did hear about it he was profoundly skeptical. Both he and Blackett argued that a radio beam from a transmitter on the continent would spread out far too much to be of use for accurate location of a bombing target by the time it reached England. But Jones saw a way it could work, with two beams aimed at slightly different angles; the receiver on the airplane could then detect the very narrow band—it could be as little as 100 yards wide even at a distance of 200 miles from the transmitters—where the two signals overlapped.

In retrospect it was clear that Tizard's mounting fury about Lindemann's takeover of the whole wartime science enterprise was clouding his judgment. Worse followed. On June 21, Jones arrived at his office to find a note on his desk: "Squadron Leader Scott-Farnie telephoned and says will you go to the Cabinet Room at 10 Downing Street." Jones was, of course, certain that this was retaliation from one of his officemates for one of his own many practical jokes. But a few phone calls quickly convinced him otherwise; he leapt into a taxi and arrived twenty-five minutes into a meeting Churchill had summoned to consider the beam question. Tizard sat on one side of the room, Lindemann on the other; Jones tactfully ignored Lindemann's gesture to take a place next to him and found a spot on neutral ground at the end of the table.

From the discussion taking place Jones realized that none of the august personages assembled had quite grasped the matter, so when Churchill at one point called on him to explain a technical point that had arisen, Jones courageously seized the moment. "Would it help, Sir," he asked, "if I told you the story right from the start?" The prime minister was taken aback for a second but quickly replied, "Well yes it would!" Churchill would later write with unabashed admiration of the young "Mr. Jones . . . unrolling his chain of circumstantial evidence the like of which for its convincing fascination was never surpassed even by the tales of Sherlock Holmes or Monsieur Lecoq."[21] The prime minister asked what should be done next; Jones replied that an aircraft should be sent up to try to detect the beams. The prime minister so ordered. The next day the hard evidence was in hand, and work began at once to devise a series of successful jammers and countermeasures to throw off the German system.

Tizard had strongly opposed Jones and Lindemann at the critical cabinet meeting, a fact Churchill never forgot. "If we had listened to Sir Henry Tizard in 1940," he remarked on one occasion two years later, "we should not have known about the beams."[22] Recognizing the hopelessness of his own position, and the now total eclipse of the influence of the CSSAW by Lindemann, Tizard immediately wrote out his resignation as both chairman of the committee and scientific adviser to the Air Ministry. It had become impossible, he explained, to continue when he no longer had the confidence of the prime minister, and his authority and responsibility was being undercut all the time. Privately he wrote Hill: "The fact is that Winston is trying not only to be Prime Minister and Commander-in-Chief, but also, through his pets, to control all the scientific work of the Department. . . . If this goes on we are bound to lose the war." Blackett and Hill both resigned from the committee as well; Tizard recommended the CSSAW be disbanded given that all of its effective power had been absorbed by Lindemann.

Dowding wrote to say he was "very resentful" of Tizard's treatment but understood: "The present witch doctor is firmly established for the time being."[23]

SINCE THE WAR BEGAN Tizard had been thinking about ways to cultivate ties with American scientists. "Bringing American scientists into the war before their government" was how he privately described it. At Tizard's suggestion Hill was sent to the British embassy in Washington in February 1940 to make contact with the scientists he knew in the United States, and to discuss possible areas of cooperation; officially Hill was an assistant air attaché though the Royal Society continued to pay his salary and the Air Ministry covered only his expenses.

Hill quickly concluded that an exchange of technical information would benefit both sides. It was in British interests to help American defense preparations by sharing their technical knowledge; likewise there was much that

American science and industry could do to help Britain. The main obstacle was British standoffishness, plus the usual obsession with secrecy in the military services. "Our impudent assumption of superiority, and a failure to appreciate the easy terms on which closer American collaboration could be secured, may help to lose us the war," he complained:

> My main thesis is that *we could get much more help in the U.S. and Canada if we were not so damnably sticky and unimaginative.* There is an intense eagerness to help which we do not exploit: (a) because we are such bloody asses, or (b) because we are so sure we can win without anybody's help.

Hill had returned to London in June and joined Tizard in an appeal that Churchill propose directly to the president of the United States a formal exchange of technical information. Their key recommendation was that no strings be attached: "The essential point of this is that we should offer any information desired, *without condition*, since we realise that America is fundamentally engaged in the same struggle for civilization as we are."[24]

Tizard's campaign to continue and expand these scientific exchanges with the Americans found a natural ally in the new prime minister. Churchill, too, had been hatching all manner of schemes to nudge the United States off its isolationist fence-sitting from the moment he became prime minister. The most visible in the summer of 1940 was his request to the United States for the loan of fifty First World War–era destroyers to help make up Britain's dire shortfall in escort vessels. The need was real enough, but Churchill was arguably more interested in the symbolic value, and the precedent it would set. As he candidly told the House of Commons, the destroyer deal would ensure that the two countries would henceforth be "somewhat mixed up together." More cooperation would inevitably follow, all to the good. "I could not stop it if I wished," he declared. "No one can stop it. Like the Mississippi, it just keeps rolling along. Let it roll on. Let it roll on—full flood, inexorable, irresistible, benignant, to broader lands and better days."[25] Behind the scenes, Churchill was prodding his military chiefs to share intelligence and technical data with their counterparts in the U.S. Army and Navy on a scale unheard of before, to exactly the same end.

So he needed little convincing to approve Tizard's new proposal to open a direct channel between Britain's and America's leading scientists. In August, Tizard had a personal interview with the prime minister and came away with Churchill's blessing for him to travel to the United States and offer "all the assistance I can on behalf of the British Government to enable the armed forces of the U.S.A. to reach the highest level of technical efficiency." Tizard told the other members of what was now officially known as the British Technical and Scientific Mission to the United States that their main object was to create goodwill, not strike deals.

With typically comical penny-pinching, the Treasury then informed Tizard that the government would be unable to pay his out-of-pocket expenses for the trip; regulations permitted paying "outside people" only when "traveling for short periods in this country on official business." He was booked on a flying boat whose passenger seats had been stripped out and whose captain had never crossed the Atlantic before. Tizard tried to take the discomforts in stride, content with the knowledge that he was embarked on a mission of unprecedented importance. The delegation carried with them a large black-enameled metal trunk, purchased at London's posh Army & Navy Stores, and it was filled with some of the most priceless fruits of British military research. The prize was something called the cavity magnetron, developed by physicists at the University of Birmingham; it was a revolutionary, highly compact device that generated powerful radar signals at unprecedentedly short wavelengths, centimeters rather than meters. It opened up the possibility of installing radar sets on aircraft, something almost unimaginable with the standard metric-wave radar transmitters, which required correspondingly sized aerials, and amplifiers filled with power supplies and vacuum tubes weighing hundreds of pounds.

Tizard reached Washington on August 22, 1940, and four days later was taken by the British ambassador, Lord Lothian, to see President Roosevelt at the White House. They slipped in the back door to avoid the camera crews and had a cordial meeting; Roosevelt "talked generalities" but Tizard fell under the FDR charm as so many others did. "A most attractive personality," he noted. The American military officers were a bit warier in their meetings with Tizard but the magnetron proved to be the icebreaker, convincing the Americans that "we were putting all our cards on the table," as the physicist John Cockcroft, who had joined the mission, put it. "From then on we had no difficulties." American scientists had been independently working on centimeter-wave radar and had built a 10-watt transmitter using a vacuum tube called the klystron. The cavity magnetron instantly increased the power that could be generated by a factor of 1,000. It was, in the words of the official historian of the U.S. wartime research organization, probably "the most valuable cargo ever brought to our shores."[26]

The American scientific leadership needed no convincing; they fully shared Tizard's agenda as their own. That was especially true of Vannevar Bush, who had come to Washington as the war approached with one specific intention in mind: to create a national scientific research organization that could focus America's considerable civilian scientific talents on military needs, and to ensure that the military used them. To get himself to Washington, Bush had declined an offer of the presidency of MIT, accepting instead the top position at the Carnegie Institution, which the industrialist and philanthropist Andrew Carnegie had established in Washington in 1901 to support basic scientific discovery. Bush was an electrical engineer from an old New England family; to colleagues at MIT who came from elsewhere in the country he seemed the quintessential Yankee. His characteristic pose was leaning back in his chair, feet on his desk, "interspersing puffs of smoke from his eternal pipe with bits of dry humor or laconic wisdom, spoken in his Yankee twang," in the recollections of one colleague.[27]

The image of the droll Yankee cracker-barrel philosopher was a bit misleading. Bush was a brilliant engineer, having designed and built at MIT the first calculating machine that could solve differential equations. He was also a masterful and instinctive politician. In Washington he seemed to know everyone. When the war began he brought together a group of the country's recognized scientific leaders to press his plan. They were Frank Jewett, president of Bell Telephone Laboratories, who was also president of the National Academy of Sciences; James Bryant Conant, a chemist and the president of Harvard University; Richard Tolman, a professor of physical chemistry and mathematical physics and dean of graduate studies at Caltech; and Karl T. Compton, the president of MIT. As Bush recounted:

> We were agreed that the war was bound to break out into an intense struggle, that America was sure to get into it in one way or another sooner or later, that it would be a highly technical struggle, that we were by no means prepared in this regard, and finally and most importantly, that the military system as it existed, and as it had operated during the First World War, which we all remembered, would never fully produce the new instrumentalities which we would certainly need.[28]

Bush drafted a plan for an organization he called the National Defense Research Committee. It would come directly under the president of the

United States, rather than the military services, and have its own source of funding. In early June 1940 he made the rounds of official Washington selling the idea. Bush understood enough about the political scene to know that the way to get to the man who really mattered, President Roosevelt, was through his aide Harry Hopkins. Short, intense, chain-smoking, devotedly loyal to his boss, Hopkins was a source of enduring resentment and jealousy among other members of the White House inner circle for his close relationship with the president. He had been with Roosevelt since FDR's days as governor of New York, had been administrator of the Works Progress Administration early in the New Deal, and Roosevelt valued him for his fearless ability to cut through existing lines of authority and get things done. Roosevelt once explained to his defeated Republican opponent Wendell Willkie why he kept Hopkins around, despite all the ill will he generated: "Someday you may well be sitting here where I am now. And when you are, you'll be looking at that door over there and knowing that practically everybody who walks through it wants something out of you. You'll learn what a lonely job this is, and you'll discover the need for somebody like Harry Hopkins, who asks for nothing except to serve you."

Bush and Hopkins were almost exact opposites in personality, politics, even looks. But as Bush recalled, "something meshed" when the two men met. On June 12 Hopkins took him in to see the president. Bush had with him his plan for the NDRC "in four short paragraphs in the middle of a sheet of paper." He left with Roosevelt's "OK—FDR" at the bottom of the sheet. "And all the wheels began to turn."[29]

As chairman of the new organization, Bush met Tizard on September 27 to work out ways to expand the exchange of scientific and technical information between the two countries. Meanwhile in London the charm offensive was continuing as British officials turned over an avalanche of technical information to an astonished delegation of visiting American admirals and generals—a "gold mine," reported Brigadier General George V. Strong in a cable to General George C. Marshall, the army chief of staff. At one point during the visit Rear Admiral Robert L. Ghormley, who was the assistant chief of naval operations, received a cabled request from the Navy Department in Washington to discreetly probe the British Admiralty for any information about Japanese defenses in the Mandated Islands in the Pacific. Ghormley inquired and received a polite but noncommittal reply from his British counterpart, who said he would see what he could do. That afternoon a motorcade pulled up, sirens screaming, at the American embassy in Grosvenor Square, messengers escorted by armed guards marched in, and, Ghormley's astonished aide recalled, "plunked down on my desk the entire portfolio of the Far East from the British Admiralty and asked me to sign a receipt for it."

Not long after, General Strong tentatively suggested an exchange of cryptologic information about German and Japanese cipher systems. This time it was the turn of the British to be astonished, but within a few months Churchill had personally signed off on the proposal and a team of American code breakers was in Britain to deliver a copy of the Japanese Purple cipher machine they had reproduced, and to receive in return briefings on British progress against the German Enigma. Although continuing British reluctance to share all of the important details about the Enigma with the Americans would become a point of serious friction between the two allies in 1942, this initial exchange of cryptologic secrets was nonetheless an unprecedented step: each side was broaching a wall of previously inviolable secrecy surrounding its most closely guarded intelligence work.[30]

The only greater secret of the war would also soon be a matter of intense British-American collaboration. On March 19, 1940, Tizard had received a memorandum from the German refugee physicist Otto Frisch and his collaborator Rudolf Peierls, another German émigré who had ended up with Frisch at the University of Birmingham, reporting their calculations of the critical mass of uranium necessary to sustain a chain reaction. Tizard gave a subcommittee of the CSSAW the task of assessing "the possibilities of producing atomic bombs during this war." Blackett was a member, as was Cockcroft and several other leading physicists; the chairman was G. P. Thomson, J. J. Thomson's son and now an established physicist in his own right. Meanwhile the NDRC, as one of its first official acts, had allocated $40,000 in July 1940 for an exploratory investigation to determine if a uranium bomb was feasible. Bush was hoping to prove it was not, which would mean there was no need to fear the Germans would make one. Tizard, too, thought that the "probability of anything of real military significance is very low."[31]

Blackett would be the sole member of the British committee, when it issued its findings a year later in July 1941, to dissent from the view that Britain should try to develop a bomb on its own. He thought the committee's estimate that they could complete the job in two years at a cost of £5 million unrealistically optimistic, and instead urged that Britain discuss joining a

U.S. program. But he did not disagree with the fundamental scientific conclusion of the report, which would prove pivotal in convincing skeptical American officials that an atomic bomb was now indeed feasible:

> We have now reached the conclusion that it will be possible to make an effective uranium bomb which, containing some 25 lb of active material, would be the equivalent as regards destructive effect to 1,800 tons of T.N.T. . . . Even if the war should end before the bombs are ready the effort would not be wasted, except in the unlikely event of complete disarmament, since no nation would care to risk being caught without a weapon of such destructive capabilities.[32]

THE LUFTWAFFE BRIEFLY came close to toppling Britain's last line of defense at the height of the Battle of Britain, in mid-August 1940. Concentrating their daylight attacks on the RAF's airfields, sector stations, and radar installations, the Germans cost Fighter Command 440 planes and a quarter of its pilots in the month of August. In retaliation for nighttime raids against British cities, the RAF carried out two small night attacks on Berlin in late August. They did negligible damage but left Berliners stunned and indignant, William Shirer recorded in his diary. Goebbels at first gave instructions to the press to play down the attacks, but after the second raid ordered a different tack. Most of the Berlin newspapers carried the same headline that day: COWARDLY BRITISH ATTACK.[33]

On September 7, Göring sent a thousand planes, the Luftwaffe's largest daylight raid yet, to strike the British capital. Air Vice Marshal Keith Park, the commander of the RAF's No. 11 Group, responsible for the crucial sector in southeast England, had a single reaction: "Thank God." The Luftwaffe's shift from the main objective of wearing down Britain's fighter forces and radar chain to striking cities to assuage a surge of retaliatory anger was a monumental strategic blunder, giving Fighter Command a desperately needed respite to reconstitute damaged airfields and radar stations, and build back up its dangerously depleted inventory of planes and trained pilots. A week later another huge aerial armada struck London. By the end of September the threat of a cross-Channel invasion eased as it became apparent that Hitler's attempt to destroy the British air force had failed. In its place Britain and Germany settled into a brutish exchange of increasingly heavy attacks on each other's cities by night. Forty thousand British civilians would be killed, 50,000 seriously injured, over the next eight months. Any remaining illusion that this would be a war fought according to the old rules that recognized such a thing as noncombatants was gone.

A few weeks before the night raids began in earnest in September 1940, General Sir Frederick Pile, commander-in-chief of the army's Anti-Aircraft Command, had asked A. V. Hill if he could recommend a scientist to help him improve the dismal performance of his gun batteries trying to shoot down enemy bombers during their night raids over England. Hill's own statistical work on antiaircraft fire in World War I was one of the earliest forerunners of operational research. "Pile instantly recognized what he needed—the quick intuition of a freshman," Hill recalled. The general had already met Blackett—accounts differ, but it may have been at a meeting of the Royal Society in early August, where Blackett delivered a paper—and been impressed by him. "Why should I not have that chap Blackett?" Pile asked Hill. Hill replied, "Why not?" Without delay, Blackett was named scientific adviser to Pile at AA Command, which shared headquarters with RAF Fighter Command at Stanmore.[34]

Antiaircraft guns were almost as old as aircraft but had always been woefully inaccurate, especially against high-flying bombers. In general, the best that gunners could hope to do was to throw so dense a wall of shells into the air that a plane was bound to run into one sooner or later. Blackett arrived at Anti-Aircraft Command in August just as the first radars were being delivered to antiaircraft artillery batteries around London. The idea of using radar data to calculate the proper bearing and elevation of the guns—the process known as "gun laying," GL for short—was eminently sensible. The only problem, as Blackett quickly found, was that no one had bothered to figure out how to connect the gun-laying radars to the guns:

> Immense scientific and technical brilliance had gone into the rapid design and manufacture of the GL [radar] sets . . . but unfortunately, partly through a shortage of scientific and technical personnel but also partly through a certain lack of imaginative insight into operational realities, hardly any detailed attention had been paid to how actually to use the GL data to direct the guns. . . . Thus the first months of the A.A. battle against the night bomber were fought with highly developed radar sets and guns, but with the crudest and most improvised links between them, belonging technically to the level of the First rather than the Second World War.[35]

General Pile was more than willing to give Blackett a free hand to see if he could improve the situation. Pile was not a natural intellectual; the eldest son of an Anglo-Irish baronet, he had spent his entire adult life in the army, winning the Military Cross and DSO as an artillery officer in France in the First World War. But he was always interested in new things. On the advice of the strategist and armored warfare theorist Colonel J. F. C. Fuller he had transferred to the Royal Tank Corps in 1923. "They like bright ideas there," Fuller told him. What impressed Pile now about Blackett was that not only was he "a first-class scientist," but he understood the realities of fighting a war and the need to make do rather than pursue some theoretical state of perfection. Pile wrote in his memoirs: "Blackett was an ideal man for the job. He spoke his mind clearly, and was always ready to admit the fact that the most desirable things sometimes may be inadvisable."[36]

Blackett was able to quickly assemble a small team mainly by hiring some young scientists he already knew. The first to arrive were A. V. Hill's son David, who like his father was a physiologist studying the mechanics of muscles, and David's close friend Andrew F. Huxley—a future Nobel Prize winner in medicine and half brother to two other famous Huxleys, Julian and Aldous. Like Blackett, Huxley had what one member of the team called "an incredible mixture of theoretical and practical expertise." He could take apart one of the large mechanical "predictors" that guided the guns "like a born mechanic." A third physiologist, Leonard Bayliss, from University College, London, became Blackett's deputy. There were also two physicists, Frank Nabarro from Bristol University and Ivor Evans (Evans's name had come from the Central Register); an astronomer, Hugh Butler; an expert from the Admiralty on gunnery predictors, Arthur Porter; an army second lieutenant, G. W. Raybould, who had been a surveyor in civilian life and, as Blackett found when he happened to meet him when inspecting an antiaircraft battery in the Midlands, had already worked out his own method for converting radar data to gun bearings; and "a girl mathematician," a Miss Keast.

The group's official status was utterly vague. Blackett was called Scientific Advisor to the Commander-in-Chief, Anti-Aircraft Command, but his salary was being paid by the Ministry of Aircraft Production. As Bayliss recalled, the scientists just decided to call themselves the Anti-Aircraft Command Research Group: "Our position was established in two ways; firstly we informed the messengers to deliver to us all correspondence so addressed, secondly we had a rubber stamp cut with the letters AACRG. It is not known whether any more formal authorisation or recognition was ever obtained."[37] Unofficially, the group soon had another name that stuck, Blackett's Circus.

Aiming an AA gun, even assuming the target was flying straight and level, required a complex calculation in spherical trigonometry. The plane's height, course, and speed had to be determined and translated into an elevation angle and azimuth of the gun; at typical bomber speeds the gun needed to be aimed roughly a mile ahead of the plane's position at the instant the gun was fired so that the shell would intercept the target as it continued forward on its course.

The Sperry Predictor was a half-ton mechanical computer designed to automate the task. A spotter would track the target, turning two dials to keep the airplane centered in a sight; the predictor, employing an array of gears, motors, and dials, would then transform the rate of change in those inputs into a continuously updated gun bearing. But the whole system was designed for visual spotting by day. There was no way to feed the radar data in directly. Blackett also found that the radar data at any given moment had a significant margin of error, so that even if the predictor could somehow be fed the data, the errors would cause the aim point of the gun to zig and zag with each updated position. Over a series of radar measurements, however, the errors tended to cancel out, and a fairly accurate track of the bomber's course could be obtained. Blackett's group first worked out a paper method of plotting the radar data and smoothing it out to obtain an average track that could be used to manually work out an elevation and bearing for the battery a given number of seconds later. Later they devised a modified Sperry Predictor—dubbed the "castrated" predictor—that allowed the operators to enter the radar tracking data by turning a dial originally intended to allow for wind corrections.[38]

About 120 guns were deployed around London, in batteries of 4 each spaced to cover the entire area of the city with overlapping fields of fire. There were only enough radar sets when the Blitz began in September 1940 to equip half of the 30 batteries. Blackett argued that there would be little to lose by grouping them into 15 eight-gun batteries instead, since the batteries without radar were extremely unlikely to hit anything and thus were already leaving gaping holes in the perimeter. A further calculation he made established that the whole idea of "complete coverage" had been an illusion anyway. Though the spacing between each of the 30 batteries in the original deployment scheme corresponded to the effective range of the guns under daytime conditions against slow-flying targets, the effective coverage of the guns using radar against modern fast bombers at night was far more limited: unless a bomber was coming almost directly toward the battery, its change in bearing took place so rapidly that the paper-and-pencil methods necessary for handling the radar data could not keep up. In fact, even the original 30-battery deployment was riddled with gaps between each battery's effective field of fire. Concentrating the guns into 15 batteries hardly changed the size of the holes in the defensive screen, while greatly increasing the odds that any given gun would successfully engage a target by giving them all access to the radar data.

The scientists calculated that it was taking 20,000 shells to bring down one German bomber when they started their work. By the following summer, Blackett's methods had cut the number of "rounds per bird," as they termed it, to 4,000. One anomaly in the data which caused much head scratching was that AA batteries deployed along the coasts were bringing down enemy bombers with half the number of rounds per bird as inland sites. "All kinds of far-fetched hypotheses were considered as possible explanations of this strange result," Blackett recalled. Perhaps the radar worked better over water; perhaps the enemy bombers flew straighter or lower while coming in across the sea than they did over land. "Then suddenly the true explanation flashed into mind." It was simply that the inevitably exaggerated claims of planes shot down by each battery could be verified on land by locating the crash site, but not over sea. "This explanation should have been thought of at once," Blackett observed, "as there is plenty of evidence to show that unchecked combat claims, made in absolute good faith, are generally much too high."[39] In the absence of Blackett's explanation, disastrous changes in the deployment of the guns and radars might have been ordered on the mistaken belief that units ought to be concentrated on the coasts for greatest effectiveness.

Another bit of statistical reassurance that Blackett was able to provide was an even more elementary yet more crucial one. Pile, as Blackett would describe him in a postwar lecture to the Royal Statistical Society, was "an extremely intelligent general but he was a little flighty in his emotions and if one night he brought down six aircraft and the next night only two he would think something was wrong with the method and want to alter the methods and reprimand the crews. . . . I circulated a table of distributions round all the staff at the Command during the blitz showing that if they expected five aircraft there would be many times they would shoot down two [or] three."

He encountered mixed results throughout the war trying to teach military officers such basic realities of probability and statistics. Another time, when he tried to explain that success in most operations is the sum of many attempts whose individual probability of success may be small, he was accosted a few days later by the pleased officer. "I say, Blackett," the officer began, "I am so glad you explained to me all about probability. As soon as the war is over I am going straight to Monte Carlo and then I really will win."[40]

THROUGHOUT THE FALL of 1940 the U-boat onslaught continued. On September 20, Kapitänleutnant Prien intercepted HX 72, a convoy of forty-one ships steaming from Halifax. BdU ordered in reinforcements, and over two nights the assembled wolf pack sank eleven ships in wave after wave of relentless torpedo attacks, many delivered at point-blank range. On October 17, SC 7, a convoy from Sydney, Cape Breton, consisting of thirty-five slow freighters barely able to make 7 knots, was beset by a pack of seven U-boats that proceeded to sink twenty of them, a shocking and demoralizing toll. "The danger of Great Britain being destroyed by a swift, overwhelming blow has for the time being very greatly receded," Churchill wrote Roosevelt in early December. "In its place there is a long, gradually maturing danger, less sudden and less spectacular, but equally deadly. This mortal danger is the steady and increasing diminution of sea tonnage." He warned that "the crunch of the whole war" was the Atlantic Ocean: "The decision for 1941 lies upon the seas."[41]

In an attempt to breathe some life into the foundering anti-U-boat campaign, the War Cabinet decided to transfer operational control of RAF Coastal Command to the Admiralty. The new commander was Air Marshal Sir Philip Joubert de la Ferté, the assistant chief of the air staff, who was keenly aware of the success of radar in the Battle of Britain—and equally, the lack of any truly effective means yet for attacking U-boats from the air. He asked if Blackett could be transferred to his new command to tackle these urgent problems. In March 1941, with a salary of £1,000 a year that the Treasury deemed "appropriate to outside scientists of distinction employed in special posts," Blackett became head of the new Operational Research Section, Coastal Command. "They have stolen my magician," lamented Pile.[42]

7

The Real War

ONE OF THE FIRST SCIENTISTS Blackett brought in to work with him at Coastal Command was E. J. Williams, a thirty-eight-year-old Welshman whose fierce bushy eyebrows and short, stocky, powerful build made his colleagues agree that he looked more like a wrestler than a physicist.

Growing up in the village of Cwmsychpant in the heart of rural Wales, he had spoken Welsh at home and did not learn English until age five. His father, a master stonemason, "possessed in full degree the passion for education which is so marked a feature of the Welsh people," Blackett would later write. On the wall of the Williams home hung a sampler embroidered by his mother with the words of a Welsh proverb, *Gwell Dysg Golud,* "Better Learning Than Riches." One evening in 1919, shortly after graduating from high school, Williams saw a notice in the local newspaper announcing four open scholarships at Swansea Technical College. The examination was the next day and the last train was gone, but Williams talked his brother into driving him the fifty miles on his motorcycle. He won the scholarship. Around this time he announced to a friend that he intended to win an 1851 Exhibition fellowship just as Rutherford had, attain a doctorate in physics, and be elected a Fellow of the Royal Society while still in his thirties. All had subsequently happened. Williams spent 1933 in Copenhagen working with Niels Bohr. At Manchester and the Cavendish Laboratory he carried out a series of theoretical and experimental studies on cloud chamber tracks that documented the discrepancies between classical and quantum mechanics in atomic collisions.

Suffering, Blackett thought, from "crippling inhibitions" that prevented him from forming any truly close friendships, Williams tended to veer from shyness to argumentativeness; he had a "wildness and unexpectedness in his behavior" that would burst out in startling ways with those he got to know. He drove a car, Blackett said, "with a complete disregard of the laws of dynamics." Another friend recalled that his sense of time "was practically non-existent." Williams would not uncommonly work until three or four in the morning and then phone a friend excitedly, only to be overwhelmed with remorse upon discovering he had awakened his victim.[1]

Blackett had brought Williams to the Royal Aircraft Establishment in early 1940 to work on a magnetic detection device for submarines, an idea that had been discussed at the Tizard Committee a few months earlier. In a short time Williams built a prototype airborne sensor, and for the first test Blackett and Williams loaded it aboard an Anson patrol plane and had the pilot buzz the large metal airplane sheds at Farnborough at a few hundred feet to see if it could detect their magnetic gradient. The device worked in principle, and the advantage of magnetic detection of a submerged submarine was that magnetic fields were unaffected by water—unlike radio waves, which was why radar could not be used beneath the surface. The trouble was that magnetic gradients dropped off with distance to the fifth power, so for practical purposes a metallic object the size of a submarine could not be detected more than 200 or 300 feet away. (American scientists were at the same time developing a different approach to magnetic detection that would prove more feasible and lead to the device known as the magnetic anomaly detector, which would play a role in the fight against the U-boats later in the war.)

Under Blackett's direction at Coastal Command, Williams now began work on what would subsequently become the single most cited example of operational research applied to war. When Blackett arrived at the command in March 1941 it was just beginning to receive aircraft and weapons that had a prayer of doing actual damage to a U-boat. At the start of the war the standard antisubmarine aircraft was the obsolete Anson, a forty-two-foot-long light transport plane whose only tangible contribution to the war effort would be as a trainer for bomber crews. The Anson could carry two 100-pound bombs and had a useful range from shore of roughly 250 miles at best. "Indeed the primary role of Coastal Command was reconnaissance and squadrons engaged in anti-submarine war were mainly intended to shadow and report the presence of U-boats to base and the Naval forces in the area," a 1944 Coastal Command lecture acknowledged.[2] The situation was so desperate at the start of the war that in December 1939 the command took to sending unarmed "scarecrow" patrols of open-cockpit Tiger Moth biplane trainers and Hornet Moth touring planes over the coasts for a few months in the hopes of forcing U-boats to waste time by submerging. Coastal Command had since replaced the Ansons with Hudson, Whitley, and Wellington twin-engine medium bombers and a small number of Sunderland flying boats, and was now, in 1941, receiving the first of the promised help from America of longer range Catalina flying boats and four-engine B-24 Liberators.

Frustrations with the ineffective 100-pound antisubmarine bombs had continued. Someone finally had the wit to consult British submariners about their own experiences on the receiving end of German air attacks. An August 18, 1940, memorandum noted that even near misses from bombs failed to do any serious damage; only in the case of a direct hit on a surfaced submarine could an attacker hope to inflict a lethal blow with an air-dropped bomb. "This itself is unlikely," the memorandum noted, "as any well trained submarine will be at least at 30' when the bomb is dropped"—having initiated a crash dive as soon as the approaching aircraft was spotted.

By contrast, a series of "disastrous" German air attacks on British submarines off Norway were almost certainly due to air-dropped depth charges, "of which our submarines have almost positive evidence."[3] Because of the dangers of the aircraft being hit by the blast from its own bomb, antisubmarine bombs had to be dropped from at least several hundred feet. There were no low-level bombsights, and accuracy was poor. A depth charge, by contrast, could be dropped from a plane just skimming the surface. That month Coastal Command acquired 700 of the navy's Mark VII 450-pound depth charges, hastily adapted with fins and a rounded nose fairing for aircraft use.

Still, as the British aviation historian Alfred Price concluded, "During 1940 aircraft had caused no more than a mild harassment to enemy submarines."[4] Pressure was mounting to produce results. U-boats, mines, surface raiders, and the Luftwaffe's four-engine Focke-Wulf 200 Condor bombers, based in France, together destroyed a half million tons of shipping in March 1941. The average number of U-boats at sea in the Atlantic and Mediterranean on any one day continued to mount steadily, reaching almost twenty in April and nearly double that by summer 1941. Since the start of the war, losses to Allied and neutral shipping had consistently outstripped the pace of new construction; the 40 million tons of shipping capacity available in 1939 had been whittled down to 35 million. "My thought had rested day and night upon this awe-striking problem," Churchill later wrote. "This mortal danger to our life-lines gnawed my bowels." On March 6 he issued a "Battle of the Atlantic Directive" ordering commanders "to take the offensive against the U-boat" and began convening weekly meetings of a cabinet-level Battle of the Atlantic Committee. The transfer of Coastal Command to Admiralty control was part of the shake-up, as was the relocation of Western Approaches Command from Plymouth to Liverpool, where it would be closer to the center of action in the war between the convoys and their attackers.

Churchill worked out a budget allocating available shipping to Britain's vital needs—it reminded him, he said, of the chancellor of the exchequer's exercise of allocating government funds to the various departments—and the outlook was, he told an aide, "terrifying." The country needed 35 million tons of imports a year; if imports fell under 31 million tons, cuts in food supplies would be unavoidable. For the first few months of 1941 imports were running at an annual rate of 28 million tons. Each day Churchill received a report on the latest shipping statistics. He would write in his memoirs of this period in early March 1941, "How willingly would I have exchanged a full-scale attempt at invasion for this shapeless, measureless peril, expressed in charts, curves, and statistics!"[5]

Rationing of food was tightened. A plan for communal canteens, much like that proposed in Zuckerman's left-wing scientific manifesto, was implemented to help stretch supplies; though Churchill, characteristically, instructed his minister of food, "I hope the term 'Communal Feedings Centres' is not going to be adopted. It is an odious expression, suggestive of Communism and the workhouse. I suggest you call them 'British Restaurants.' Everybody associates the word 'restaurant' with a good meal, and they may as well have the name if they cannot get anything else."[6]

The prime ministerial attention to the U-boat problem was as usual a mixed blessing. There were signs, however, that even Churchill was beginning to weary of Lindemann's enthusiasm for clever gadgets and wonder weapons. In April his pet scientist sent the prime minister his latest brainstorm for yet another fantastic invention, this one to detect submerged submarines: an airplane or destroyer would drop a large number of tiny magnets fitted with electric lights or self-igniting gas that would glow when they stuck to a metallic object. Churchill impatiently wrote back, "This seems to be rather far-fetched. If the aeroplanes or destroyers were as close to the submarine as necessary it would surely be better to throw explosives by bomb or depth-charge."[7] With an entire war to run, Churchill was finding he had bigger problems to worry about than the Prof's clever toys.

BLACKETT HAD LONG SINCE concluded that the way scientists could really improve things was not by trying to invent new tools anyway but by figuring out how to better use the tools already in hand. He set down his philosophy in a presentation on operational research he gave to an Admiralty panel later that year. Reprinted subsequently many times, "Scientists at the Operational Level" would become a sort of Magna Carta of operational research, and the crux of it was a plea to get scientists out of the laboratory and into operational units where they could do more good:

"New weapons for old" is apt to become a very popular cry. The success of some new devices has led to a new form of escapism which runs somewhat thus—"Our present equipment doesn't work very well; training is bad, supply is poor, spare parts non-existent. Let's have an entirely new gadget!" Then comes the vision of the new gadget, springing like Aphrodite from the M.A.P. [Ministry of Aircraft Production]. . . . In general, one might conclude that relatively too much scientific effort has been expended hitherto in the production of new devices and too little in the proper use of what we have got. Thus there is a strong case for moving many of the best scientists from the technical establishments to the operational Commands, at any rate for a time.[8]

Blackett emphasized the need for scientists in this role to steer clear of "technical midwifery" so they could focus on current operational questions.

Williams jumped right into the most basic such question of all: why were the aircraft of Coastal Command sinking so few submarines? Even with the change from the 100-pound antisubmarine bomb to the 450-pound depth charge, results had scarcely improved. As of May 1941, Williams found, aircraft had sighted U-boats 200 times, carried out attacks in 130 of those cases, and definitely sunk a total of 2 U-boats—a 1 percent success rate. In most of the attacks, the U-boat had spotted the aircraft first and had time to dive by the time the aircraft arrived in position to drop its depth charges. Coastal Command's Tactical Instruction No. 12, issued January 27, 1941, was based on a straightforward calculation: it took the average time a U-boat had been out of sight at the instant of attack—about 50 seconds—multiplied that by the U-boat's dive rate of 2 feet per second, and concluded that dropping a series of depth charges set to explode at a depth of 100 to 150 feet would be most likely to catch the submerged boat. Because the U-boat could also have traveled as much as 1,000 feet forward during the same time, the orders also advised dropping a "stick" of four charges spaced horizontally at about 250-foot intervals, to try to blanket as much of the possible target area as possible.[9]

Williams had little difficulty putting his finger on the trouble once he started sifting through the data. Shooting for an average guaranteed the worst of both possible worlds. In cases where the target was at the same depth as the depth charge setting, its odds of still being on the same straight-line course last seen on the surface were almost nil: the boat would have had time to veer to the left or right, and the possible area in which it might be lurking was so vast as to be impossible to saturate with charges. Williams calculated that the average error in aim point was at least 300 feet, well beyond the 25-foot lethal radius of a 450-pound depth charge. On the other hand, the boats that had been late to react to the aircraft's approach, and were thus still close to the surface and not far from the position where they submerged, were being attacked in the right place but with depth charges set to go off far too deep to do any damage. The 250-foot stick spacing was likewise thrown away on U-boats which had been out of sight for a quarter of a minute or less, which were sure to be within a radius of 150 feet of their last observed position. In other words, targets that were at the right depth were almost certainly in the wrong place; targets that were in the right place were *definitely* at the wrong depth. Williams summarized this conclusion at the very top of the first page of his report:

> In as many as about 40% of all attacks the U-Boat was either visible at the instant of attack or had been out of sight for less than 1/4'. It is estimated from such statistics, and the rate at which the uncertainty in the position of a U-boat grows with time of submersion, that the U-boat which is partly visible or just submerged is about ten times more important a target, potentially, than the U-boat which has been out of sight for more than about 1/4'. The very small percentage of U-boats seriously damaged or sunk in

past attacks is probably largely the result of too much attention having been given to the long submerged U-boat.[10]

There was also a satisfying calculation Williams was able to produce which showed that the actual successes attained to date agreed almost perfectly with what would be expected from following Instruction No. 12—which gave additional confidence to his conclusion that the *only* thing wrong with what Coastal Command was doing was its tactical procedures; there were no equipment malfunctions or other problems to blame for the poor performance.

Williams calculated that changing the depth charge setting to 25 feet and substantially narrowing the stick spacing would increase the percentage of U-boats sunk by an astonishing factor of *ten*—from the current 1 or 2 percent to 10 to 20 percent.[11] His typewritten report, Coastal Command ORS No. 142, was "a paper which can certainly be taken as one of the classics of Operational Research Literature," wrote C. H. Waddington, who would later join the group. It was formally submitted on September 11, 1941, and its results were incorporated almost verbatim into a new Coastal Command Tactical Instruction No. 18 issued on December 15:

> Since it is now the intention to concentrate Coastal Command effort on attacking submarines which have not had time to submerge or have not been submerged for more than 15 seconds the weapon must be made to explode so that its lethal effect will reach the pressure hull of a U/Boat which is on or not more than 30 feet below the surface.

The instructions noted that new pistols that provided a 25-foot setting were being produced for both the Mk VII 420-lb. depth charge and the Mk VIII 250-lb. version. A full stick of depth charges was to be dropped in the initial attack; larger aircraft, such as Sunderlands and Catalinas that could carry 8, were to drop them at 45-foot intervals; Hudsons, which could carry only 4, were permitted to stretch that spacing to 60 feet. In either case, though, that was a mere fraction of the 250-foot spacing specified in the earlier instructions.[12]

In one sense, Blackett said, Williams's discovery was little more than "an example of the old military precept to concentrate offensive effort on the good targets and ignore the poor targets." But it had taken the "critical but sympathetic analysis" of an outsider to spot the flaw in existing operational

The case for shallow attack

E. J. Williams calculated that air attacks would be far more effective if concentrated on U-boats that were caught on the surface or within 15 seconds of diving, before they could evasively change course. Adjusting the depth setting and spacing of depth charges led to a dramatic increase in kills.

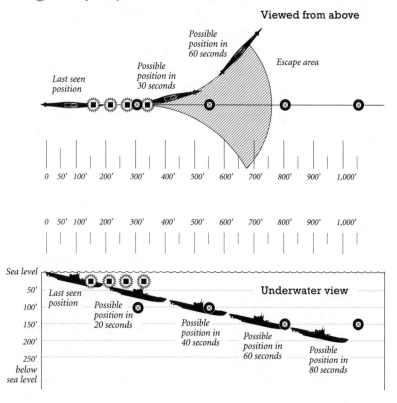

Depth-charge pattern, as originally ordered

Revised pattern for shallow attack

Viewed from above

Possible position in 60 seconds

Escape area

Last seen position

Possible position in 30 seconds

0   50' 100'   200'   300'   400'   500'   600'   700'   800'   900'   1,000'

0   50' 100'   200'   300'   400'   500'   600'   700'   800'   900'   1,000'

Sea level

50'   Last seen position

100'   Possible position in 20 seconds

Underwater view

150'   Possible position in 40 seconds

200'   Possible position in 60 seconds

250' below sea level   Possible position in 80 seconds

orders.[13] It also neatly affirmed Blackett's dictum about the relative worth of new gadgets vis-à-vis the better use of existing weapons, since Williams also was able to show that a standard depth charge with a 25-foot depth setting was almost as effective as—and much less complex than—the proximity-fused depth charge detonator he himself had been working to develop the previous year when he was at the Royal Aircraft Establishment.

BLACKETT WAS MEANWHILE producing his own piece of operational prestidigitation. In April 1941, a month after starting work at Coastal Command, he paid a visit to the operations room of Western Approaches Command in Liverpool, where a large wall map displayed the current estimated positions of all U-boats in the Atlantic. Blackett knew the number of hours being flown by Coastal Command aircraft and the areas they were patrolling. "I calculated in a few lines of arithmetic on the back of an envelope the number of U-boats which should have been sighted by the aircraft," given the actual number of U-boats operating in the area as shown on the wall of the Western Approaches Command. The theoretical number Blackett obtained from his quick calculation was four times the actual number of sightings that Coastal Command air patrols were reporting. "This discrepancy," Blackett continued, "could be explained either by assuming the U-boats cruised submerged or by assuming that they cruised on the surface and in about four cases out of five saw the aircraft and dived before being seen by the aircraft. Since U-boat prisoners asserted that U-boats seldom submerged except when aircraft were sighted, the second explanation was probably correct." All of the obvious solutions were recommended: equipping the aircrews with better binoculars, avoiding flying into the sun, improving training. Then, discussing the problem one day, an RAF wing commander asked Blackett, "What color are Coastal aircraft?"

They were in fact mainly black, as they were mostly night bombers diverted from Bomber Command. Night bombers were painted black to reflect as little light as possible from searchlights. But by day, under most conditions of cloud and sun, an aircraft is seen as a dark object against a light sky. Tests were quickly ordered and it was verified that repainting the aircraft white reduced by a fifth the average maximum distance at which the planes could be seen. The undersurfaces of the wings were the part of the aircraft that stood out in particular contrast to the sky, and a scheme of using glossy reflective white paint for these surfaces was adopted. Williams calculated that the change to white camouflage would increase the number of U-boat sightings, and sinkings, by 30 percent. The plan was implemented within a few months. Air patrols during the winter had been yielding one U-boat sighting per every 700 hours of flying. By summer 1941, with the camouflage change and other improvements, the yield had doubled to one sighting per 350 hours.[14]

It would take longer for the benefit of Williams's finding about depth settings to be realized; the depth charges, it turned out, tended to entrain a cavity of air as they struck the surface of the water, which prevented the pistol from immediately sensing the water pressure and it proved difficult to get them to go off at a depth as shallow as 25 feet. Once the problem was solved the results were what Williams's calculation predicted: by 1943, the chance of killing a submarine in an air attack with depth charges had increased from 1 to 2 percent to almost exactly 10 percent.[15]

THE ARCHETYPAL DEMONSTRATION of operational research was provided in a postwar textbook by Philip Morse, an American physicist who would follow Blackett's pioneering lead upon America's entry into the war. Morse told the story of one operational research scientist who, on his first day at a field command, noticed that there was a long line of soldiers waiting to wash their mess kits after each meal. There were four tubs available; two were for washing and two for rinsing. Observing that it took each soldier three times as long to wash his kit as to rinse it, he suggested changing the setup so that three tubs were allocated for washing and one for rinsing. The change was made. The line did not merely shrink: it vanished altogether.

It was, said Morse, a "trivial" but perfect illustration. The solution in retrospect seemed almost absurdly simple. It did not involve any new equipment. And the consequences of the change were what mathematicians call nonlinear—a small perturbation produced disproportionately large results. That was a common feature of processes in which a flow hits a bottleneck. Delays and obstructions feed on themselves: the longer a line gets, the longer it tends to get. Conversely, eliminating even a small obstruction can cause a backup to disappear altogether.[16]

Blackett made some similar observations in lectures in which he looked back on the birth of operational research in the war and the lessons learned:

Quite a number of these results were really quite simple, and appear even more so when talked about now. There is now a kind of deceptive simplicity about the results of these investigations which tends to make previous tactics seem rather stupid and ill-advised. Actually, it wasn't like that.

Nine times out of ten, he said, the conventional way of doing things turned out to be best. And although it often looked superficially as a matter of "the bright scientist suddenly intervening and telling the experts what to do," it was much more a matter of scientists asking the right question—and only after having thoroughly "soaked themselves" in the problems of the operational command by attending staff meetings, looking at intelligence reports, hearing orders given, and seeing how things worked.

He added a crucial observation about sociology: the only way to get "very busy and harassed officers" to pay attention to operational research findings was "to concentrate on results which appeared certain . . . the scientists had to get results so definite that they didn't need a mass of statistics to prove them correct to the Service people." Statistics certainly was a frequent tool of the analysis that pointed out to the researchers what changes were likely to produce results, but the results themselves had to be so dramatic as to speak for themselves. That meant, too, that a scientist "put into this kind of administrative environment" had to behave "in quite a different way from in his own laboratory," Blackett said: "His job is to improve matters if he can, and, if he cannot, to say nothing."[17]

Operational scientists had to have a basic grasp of mathematics and probability theory, but beyond that Blackett thought the best candidates "knew nothing about the subject beforehand" and thus would be "prepared to ask questions that more 'instructed' people might not think of asking."[18] In his original version of "Scientists at the Operational Level," Blackett described in detail the talents required and the participatory role that the operational research staff needed to play to be effective:

A considerable fraction of the Staff of an O.R.S. should be of the very highest standing in science, and many of them should be drawn from those who have had experience in the Service Technical Establishments. Others should be chosen for analytic ability, e.g. gifted mathematicians, lawyers, chess players. An O.R.S. which contents itself with the routine preparation of statistical reports and narratives will be of very limited value. The atmosphere required is that of a first class pure scientific research institution, and the calibre of the personnel should match this. All members of an O.R.S. should spend part of their time at operational stations.[19]

The physical chemist E. C. Baughan, who joined Coastal Command ORS later that year after returning from Princeton (where he had been a visiting professor when the war broke out) catalogued the first members of Blackett's staff there, and if there were no lawyers or chess players, they were indeed an eclectic bunch: "Three physicists (and one physical chemist), three communications experts (one Australian), four mathematicians, two astronomers (both Canadian), and about eight physiologists and biologists, including an expert on the sex life of the oyster." He added: "It was not clear which background was the best." A subsequent list would have included a classical archaeologist, several economists and statisticians, and one botanist. John C. Kendrew, who would win the Nobel Prize in chemistry in 1962, was an early member of the group; C. H. Waddington joined the following year and served as its head from 1944 to mid-1945. Other famous members of the Coastal Command group would include the mathematician J. H. C. Whitehead and the geneticist Cecil Gordon.

Leonard Bayliss thought that the large number of physiologists and biologists in the operational research sections was not a coincidence. It was true that most physicists had already been siphoned up for radar work (and some would later be drawn into the Manhattan Project), which partly explained the predominance of scientists from other fields in the operational research sections. But Bayliss thought "biologists were more accustomed to making the best of very imperfect and inadequate data, and drawing some sort of conclusions from them; physicists would be more likely to throw in their hands until the apparatus was improved and adequate data was provided." Baughan recalled one of the mathematicians in the group sarcastically proposing a definition of operational research as "an Experimental Science where a number is equal to its square root." It certainly required a tolerance for uncertainty and approximation. But Baughan noted that many of the most dramatic results the scientists produced were based on calculations that employed extremely rough estimates of the underlying data.[20]

Scientists were by nature mavericks when it came to authority, and a willingness to plunge through accepted dogma and maintain independence was a part of the job that came naturally to most. Andrew Huxley remembered Blackett advising him that it was better not to accept a commission as an officer since "the great advantage of being a civilian was that you could answer back to the senior officers."[21] As T. E. Easterfield, a New Zealander fresh out of Cambridge with a Ph.D. in pure mathematics who joined the Coastal Command ORS in 1942, recalled, being a civilian also meant that the scientists could go around and talk to NCOs and officers alike on an informal basis, and get the real story. Indeed, establishing good working relations throughout the command was an essential part of getting anything done, the scientists quickly discovered. Easterfield recalled that Harold Larnder, who took over as director of Coastal Command ORS in 1943, had a "marvelous gift for hob-nobbing with senior officers, picking up what was in the air, and spotting problems before they were formulated by the officers themselves." Easterfield added: "We lived at the HQ of the Coastal Command, ate in the same mess, and drank in the same bar. For some people (not me), the latter provided a very big part of the contact." Easterfield recalled one of his colleagues declaring, "Ninety percent of operational research is beer!"[22]

RITCHIE CALDER, the science correspondent for the *Daily Herald*, was a member of the Tots and Quots and had gotten to know many of the operational research scientists. In the 1940s it was still possible to publish a newspaper that appealed to a large working-class readership with serious articles about left-wing politics, literature, economics, and science; the *Daily Herald* had been a hard-left antiwar voice in the First World War, had supported the Russian Revolution, and was later owned by the Trades Union Congress. George Lansbury, the pacifist Labour Party leader of the 1930s, was its editor during and after the First World War; Siegfried Sassoon was literary editor. (In the 1960s the paper would be bought by Rupert Murdoch, renamed *The Sun*, and continue its mass appeal as a tabloid best known for featuring on page 3 each day a photograph of an attractive and stark-naked young woman.)

One day in 1941 Calder dropped in on Cecil Gordon, a geneticist he knew at the University of Aberdeen. Gordon was busy counting the hairs on the antennae of fruit flies searching for mutations he had been hoping to induce. The next time Calder heard of him, Gordon was at Coastal Command working out the mathematics of aircraft flying and maintenance schedules. Calder was struck by the same scientific detachment he brought to both tasks. "Gordon," he said, "treated Coastal Command as though it were a colony of his pet *drosophila*."[23]

Keeping a certain emotional distance from one's subject was scientific habit; when the subject was the lives and deaths of thousands of men it was also probably a necessity. By 1941, the war at sea was already driving the men who knew its cruel realities firsthand to the breaking point. The monotony and constant strain of station keeping in a convoy of merchant ships stretched out in ordered lines across miles of ocean was relieved only by the terror of a torpedo exploding without warning in their midst when a U-boat broke through the always overstretched escort screen. Many of the crewmen were raw, inexperienced, and not infrequently terrified. They were quickly disabused of any ideas they might have had about the romance of life at sea; merchant ships had none of the spit and polish and ritual of the regular navy that at least gave a pretense of esprit de corps and a glimmer of a nobler age. An American seaman fresh out of merchant marine school remembered reporting for his first watch at sea. Approaching the chief mate on the bridge he saluted briskly and barked out, "Relieving the watch, Sir!" That was what he had been taught to do at school. The chief mate stared at him dumbfounded for a moment before muttering, "Oh my, how lovely."[24]

The more experienced hands offered some cynical comfort to their new messmates about what they could do to increase their odds of surviving a U-boat attack. On a ship carrying a heavy cargo like iron ore or steel plate, you slept on deck because the ship would sink like a rock if it were torpedoed and you had only seconds to scramble overboard. On a ship carrying a lighter load you slept belowdecks but left your clothes on and the door open to give yourself a chance of getting out quickly. If you were aboard a tanker, or a freighter loaded with ammunition, you got undressed, shut the door, and got a good night's sleep because you didn't have a prayer anyway if the ship were hit.

Worse was what frequently happened to the survivors who *did* manage to get out of their sinking ship in time. It did not take long to learn that stopping even briefly to try to pick up survivors just made another ship in the convoy a sitting duck; there was nothing to do but steam on. Some convoys had a rescue ship that followed astern to retrieve survivors but many did not, and in any case the Germans seemed to be deliberately targeting them; so many were torpedoed by 1942 that there were only enough available to accompany one convoy in four. The image of passing literally within feet of helpless men left to die in the dark and freezing waters of the North Atlantic was a horror few would forget. "I saw it first in HMS *Alaunia* in 1940," wrote Hal Lawrence, a sailor in the Canadian navy:

They shout, even cheer, as you approach; the red lights of their life jackets flicker when they are on the crest of a wave and are dowsed as they slip into the trough; their cries turn to incredulous despair as you glide by, unheeding, keeping a stoical face as best you can. But the cold logic of war is that these men in the water belong to a ship that has bought it and that a couple of dozen more ships survive and must be protected. . . . Each time was as bad as the first. We *never* got used to it.[25]

The small corvettes Churchill had ordered hastily into construction to fill the gaping shortage of escort vessels were, in his own words, "cheap and nasty." They weren't so cheap, but no one disputed the other part of Churchill's description. A scant 200 feet long, they carried a single 4-inch gun on the bow and racks of depth charges in the open stern. The bridge was open to the elements, too, save for a small enclosed wheelhouse and another boxlike cabin holding the asdic set. The Royal Navy's theory was that fresh air kept the watch awake and on their toes and that an enclosed bridge hindered visibility. In fact, in any bad weather standing watch was "sheer unmitigated hell," said one young Canadian officer. The ships were originally planned for a complement of 29 officers and men but that was increased to 47 and then 67, with the result that 55 enlisted men shared two 20-by-14-foot compartments, two toilets, one urinal, and three wash basins. There was no forced ventilation system and the first fifty ships that were built had no insulation either, which caused the walls to sweat with heavy condensation, causing epidemics of pneumonia and TB among their crews. In rough weather water simply slammed down through the standing ventilator pipes, flooding the mess decks and wardroom and washing up a tide of spilled food, sodden clothes, and loose gear in chaotic piles.[26]

The food was abominable. The only passage from the small galley, at the rear of the superstructure, to the bridge and the forward crew quarters was across an open well deck which was frequently swept by heavy seas, which meant meals arrived cold, if they arrived at all. The galley's small refrigerator could hold only five days' worth of fresh meat and vegetables, after which the menu settled into an invariable and dreary procession of canned sausages, canned corned beef, canned stew, hard biscuits, and tea. (The canned stew came in an ornate container labeled OLD MOTHER JAMESON'S FARM HOUSE DINNER. "I must remember *never* to go to dinner at Mrs. Jameson's," one officer sardonically remarked.)

To add insult to injury, the corvettes had all been given the names of flowers: HMS *Gladiolus*, HMS *Periwinkle*, HMS *Buttercup* . . . It almost seemed like a bad joke. The sharpest evocation of life aboard one of these small Atlantic escort vessels is found in the lightly fictionalized novel *The Cruel Sea*, by Nicholas Monsarrat, a journalist who joined the Royal Naval Volunteer Reserves in response to a call at the start of the war for gentlemen yachtsmen. In one vivid passage he captured the sheer unseaworthy ugliness of the Flower-class corvette his fictional alter ego joins as a new RNVR sublieutenant: "Broad, chunky, and graceless . . . not much more than a floating platform for depth charges . . . She would be a natural bastard in any kind of seaway, and in a full Atlantic gale she would be thrown about like a chip of wood." The ships barely made 16 knots, slower than a surfaced U-boat. On a shakedown cruise, running down to the Isle of Arran in a very moderate sea her first time out of harbor, Monsarrat's fictional HMS *Compass Rose* achieved a mind-boggling 40-degree roll.[27]

But what Monsarrat evokes most of all is the endless complications of turning tactical theory into anything approaching successful practice where human beings, with all of their vanities and frailties and endless personal crises, are involved—the entire sphere of warfare that lay beyond the realm of equations and analysis and rationality of Blackett, Gordon, and the other operational scientists. Early on, Monsarrat's Sublieutenant Lockhart realizes that everything he has read and learned and taken to heart about naval tradition, duty, procedures, and tactics is being "destroyed and poisoned" by the bastard of a lieutenant he is stuck with. The lieutenant is a bully and a shirker and an opportunist who sloughs off every responsibility he can to the junior officers and covers up for his incompetence with hectoring fault finding. The crew, for its part, is mostly able and willing but beset with the sad domestic dramas of the poor; one sailor goes AWOL for seventeen days and refuses to explain until the captain sympathetically coaxes his story out of him: a pal in his North London neighborhood had sent him a letter telling him all about his wife's increasingly notorious goings-on with a commercial traveler in his absence.[28]

As in his real life, Monsarrat's character is by the end of the war a captain in command of a destroyer and has had several daring successes in epic depth charge battles with U-boats. But even these victories seem almost to defy the larger truth of the story—that war, up close, is mostly chaos and chance and fear. Many of the Flower-class corvettes were manned by inexperienced crews of the Royal Canadian Navy, which like the British navy had disregarded the submarine threat and been woefully unprepared for escort duty. In all, Canadian shipyards would build 130 escort vessels in the course of the war; by May 1941 the RCN had enough ships available to take over full responsibility for escorting convoys from Newfoundland to a line running through 35° west longitude, where the ships would be turned over to British escorts while the Canadians would take charge of a westbound convoy for the return voyage.

The Canadian vessels, though, did not even have the advantage of modifications the Royal Navy had since made that mitigated some of the worst flaws of the original design. The Canadian crews were almost all ex–merchant mariners and hastily recruited reservists; many of the latter came from Canada's prairie provinces and literally had never seen the ocean before. "We were all badly trained, scared stiff, and most of the time wished to God we had joined the air force," said one.[29]

THERE WAS AN EERIE SYMMETRY between hunter and hunted. The Type VII Atlantic U-boats were almost exactly the same length as the corvettes that sought to find and destroy them, and carried a crew of almost identical size, living in equally deplorable conditions. It would fall to a German novelist, Lothar-Günther Buchheim, to render the most vividly impressionistic sense of life on the U-boats, just as Monsarrat had on the other side. Buchheim was a journalist, too; in 1940 he had volunteered for the Kriegsmarine and been assigned as a war correspondent in the navy's propaganda unit. The following year he was ordered to sail with *U-96* on her seventh patrol and produce a stirring story, with photographs, of the brave men of the U-boat arm in action. Thirty years later he wrote what he intended to be a far more honest and gritty account in his autobiographical novel *Das Boot*. The book opens in 1941 and already the experienced U-boat men are prematurely gray and cracking up. They have all, at least once, returned limping into port with boats whose continued seaworthiness seemed to defy the laws of physics—"upper deck demolished by aircraft bombs, conning tower rammed in, stoved-in bow, cracked pressure hull." But,

each time they did come back, they were standing bolt upright on the bridge, acting as though the whole mission had been mere routine. To act as though everything were a matter of course is part of the code. Howling and chattering of teeth are not allowed. U-boat Headquarters keeps the game going. For Headquarters, anyone who still has a neck and a head and all four extremities attached to his torso is all right. For Headquarters, you're certifiable only when you start to rave. They should long ago have sent out fresh, unscathed men to replace the old commanders in the front-line boats. But, alas, the unscathed novices with their unshaken nerves happen to be far less competent than the old commanders. . . . But that's the way it goes: U-boat Headquarters has been struck by blindness. They don't see when someone is on his last legs, or don't want to see.[30]

The U-boat men increasingly relieve the strain in total dissipation ashore. At a bacchanalia the night before their departure, the officers of the boat Buchheim's narrator is assigned to are soon far beyond inebriation, staggering back and forth between a nightclub and the nearby whorehouses of the French port town. One officer passes out in the lavatory in a pool of urine and vomit; another solemnly pours beer into the piano; another pulls out his service revolver and sends the rest of the drinkers diving under their tables as he shoots out, one by one, the classical mythological figures tastefully depicted on a mural over the orchestra platform. Gestapo officers morosely but silently watch the proceedings.

If the corvettes were little more than floating depth charge platforms, the Type VII U-boats were little more than seagoing torpedo platforms. Each boat carried twelve torpedoes, each twenty-three feet long and weighing two tons. The torpedoes were the boats' raison d'être and imperative and everything gave way to their needs—comfort, space, payload, work schedules. The foremost compartment of the boat was where most of the regular seamen bunked but its name was the torpedo room, which accurately reflected priorities. Four torpedoes were kept loaded in their firing tubes; six reloads had to be accommodated on and beneath the deckplates, taking up space in the already impossibly cramped room, barely ten feet wide and not tall enough for a man to stand up in. In what space was left over twelve bunks hung in rows, shared by twenty-four men who took turns sleeping in them in shifts—"hot bunking." The torpedoes, almost like living things, demanded constant nursing and tending. Every few days their delicate mechanisms had to be checked and adjusted, batteries charged, guidance and depth-setting systems inspected. When that happened, the bunks had to be stowed away, the deckplates pulled up, the torpedoes in their tubes manhandled partway out and the whole room became more machine shop than living space.

The officers and petty officers occupied only slightly more roomy compartments farther aft. Every compartment doubled as passageway, mess room, and workspace; sleep was constantly being interrupted by the mechanics and seamen passing to and from the engine room, control room, or bridge for their watches, by tables being folded out for mealtimes (when the lower bunks served as benches), by deckplates being wrested up to get at batteries or other equipment. The control room at the center of the boat was an incredible maze of cable chases, compressed air pipes, gauges, and dials that gave one the feeling of crawling through the innermost workings of a power plant or factory. There were two toilets for the forty-four officers and men, but the one next to the galley was inevitably given over to the more pressing needs of storage and was usually crammed with crates of vegetables, potatoes, cheese, cans of coffee, bottles of fruit juice. ("More space for food and less for shitting! You try to make sense out of it!" the bosun boisterously declares to Buchheim's narrator.) At the start of a cruise, moving anywhere meant ducking or dodging around hanging provisions: sausages and sides of bacon suspended from the ceiling of the control room, loaves of bread filling hammocks in front of the radio room.

Whenever the boat was running on the surface four men stood watch on the bridge, each assigned one quadrant to scan constantly; in even a small chop waves crashed over the deck drenching the lookouts. Still, the fresh air was a welcome relief for most; below the stench was a constant presence; it hit you like a physical force, a mixture of diesel fumes, lubricating oil, engine exhaust, damp rot, cooking odors, and the acrid and omnipresent fug of unwashed bodies, ineffectually masked with cologne. After a few weeks green mold grew on leather belts, black mildew on shirts, thick crusts of yellow fuzz covered the bread, and faces and arms were pocked with infected scabs and boils.[31]

The discomforts of day-to-day life were nothing compared to Buchheim's descriptions of experiencing a depth charge attack: hours on end of cat and mouse, lying still hundreds of feet down, the churning propellers of destroyers and corvettes growing steadily louder, followed by the relentless rain of blasts, each of which could split the pressure hull open like a can opener, with certain death to follow.

FOR THE THREE MONTHS preceding Churchill's Battle of the Atlantic Directive, British escorts had not sunk a single U-boat. That changed on March 7, 1941—the very day following the prime minister's order to "take the offensive." The corvette *Camellia*, escorting a westbound convoy south of Iceland, sighted a U-boat on the surface. A second corvette, HMS *Arbutus*, joined the attack, and over the next four hours the ships repeatedly depth-charged a spot where air bubbles were seen rising to the surface. Survivors from the U-boat—she would turn out to be *U-70*, under the command of Kapitänleutnant Joachim Matz—later told their British captors that the explosions had started a leak that sent the boat sinking out of control. The depth gauge hit 200 meters—its maximum reading—and still the boat kept dropping farther as cracking sounds rent the hull and paint began flaking off the sides of the walls. With only a small quantity of compressed air in reserve and the stern down at an angle of 45 degrees, the crew huddled in the forward compartment to try to bring her on an even keel, set the electric motors full ahead, and blew the tanks. As the boat broke the surface the crew opened the scuttling vents and spilled out of the conning tower. The British ships fished Matz and twenty-five of his officers and men out of the water; twenty others were lost.[32]

Over the next week and a half three more U-boats were sunk by escorts in depth charge and ramming attacks. Otto Kretschmer of *U-99*, Dönitz's top ace, was taken prisoner; Joachim Schepke of *U-100*, the next highest in the tonnage score, was crushed to death when the destroyer *Vanoc* sliced through her conning tower, ramming the surfaced boat as she tried to flee in the darkness; Günther Prien and *U-47* were lost either in a depth charge attack or from some other cause. The string of British successes had to be partly coincidence, but they were a harbinger that the tide of the battle was shifting at last.

Just after noon on May 7, 1941, a Royal Navy task force of three cruisers and four destroyers reached a bleak stretch of ocean northeast of Iceland, changed course to the east, and formed into seven parallel lines spaced at ten-mile intervals. For the rest of the day they steamed steadily eastward, combing a square of sea sixty-nine miles on a side, which was one degree of latitude. Harry Hinsley, the young German history student who had become Bletchley Park's authority on German naval communications, had determined that somewhere within that patch of ocean, designated on German charts as grid square AE 39, was the German weather trawler *München*. Aboard her, Hinsley fervently hoped, were the maddeningly elusive pieces of the puzzle that would allow the Bletchley code breakers to crack the naval Enigma once and for all.

In the fading afternoon light a lookout aboard the cruiser *Edinburgh* spotted smoke on the starboard bow. After a short chase two of the destroyers overhauled the German trawler and sent a few shells hurtling into the water near her; the crew promptly abandoned ship and a boarding party from the destroyer *Somali*, which had meanwhile raced alongside at top speed, leapt aboard. They were soon joined by an officer from the *Edinburgh*: Jasper Haines, a captain in the Naval Intelligence Division. He seemed to know exactly what he was looking for, the other members of the boarding party later recalled, making his way directly belowships and returning a few minutes later with a bundle of papers in his hand. Within a few hours Haines was heading for Scapa aboard the destroyer *Nestor*, with orders to get to Bletchley as soon as possible with his finds.[33]

Two days later an unplanned bonus was added to Hinsley's haul. The destroyer *Bulldog* was escorting a westbound convoy of forty ships when two enormous torpedo blasts suddenly erupted in the bright noon. The *Bulldog*'s captain, Commander A. J. Baker-Cresswell, sheered off from the head of the column with two corvettes to chase down the attacker. Almost at once the corvette *Aubretia* made asdic contact and began dropping depth charges. Only moments after having turned and dropped a second pattern of depth charges, the escorts were rewarded with "the dream of all escort vessels," as Sublieutenant David Balme of the *Bulldog* put it, the sight of a U-boat blown to the surface. Thinking quickly, Baker-Cresswell saw the chance to capture the boat before her crew could scuttle her, which was the standard German practice. He ordered his crew to open fire with every available weapon. A hail of shells and bullets from the ship's two 4.7-inch guns, machine guns, even an antiaircraft gun clattered in deafening syncopation against the metal hull of the U-boat. The German crew began spilling out onto the deck and into the water.

Baker-Cresswell turned to Balme. "Right, we will board her. Sub, you take this sea boat." Balme was twenty years old, the sea boat was a rowboat propelled by five oarsmen, and the sea was running with six-foot rollers. Reaching the U-boat, Balme climbed into the conning tower, waved his revolver around a bit wondering apprehensively if any Germans were still down there, and plunged down the ladder. The boat was deserted. Balme formed the rest of the boarding party into a human chain and passed charts, books, papers, and gear up the ladder and into the sea boat. The captured boat, *U-110*, was taken under tow by *Bulldog* but sank the next morning ("one of the greatest blessings in disguise," Balme later understood, as it kept the Germans from learning that the British had been able to retrieve the critical documents from the boat).[34]

On May 13 a Royal Navy intelligence officer met the returning ship at Scapa Flow, and that same day he was on a plane for London, and then on to Bletchley Park, where he arrived that afternoon. That night at 9:37 the teleprinters at the Admiralty's bombproof concrete "Citadel" in Whitehall came rattling to life with the start of an uninterrupted stream of a hundred deciphered naval Enigma messages from the Bletchley code breakers.[35]

THE BREAKTHROUGH HAD COME from a list of daily settings for the naval Enigma network taken from the *München* and, even more priceless, the external code tables used to encrypt the eight-letter indicators which specified at the start of each message the precise starting position of the rotors used for the machine for that particular message. Those tables were among the haul of papers from *U-110*. (Balme had also retrieved an intact Enigma machine, but that added nothing the code breakers had not already long had in hand.)

Since the previous summer the Bletchley cryptanalysts had been breaking army and air force Enigma traffic using the first of the behemothlike mechanical calculators they called the "bombes." Built by the British Tabulating Machine Company, they cost £5,000 apiece, weighed one ton, emitted a perpetual stream of leaking machine oil, and were plagued with temperamental wire contacts on the dozens of spinning rotors which reproduced the Enigma machine's wiring. Each little wire brush contact had to be preened with tweezers before each run of the machine to keep them from shorting.[36]

They were a stroke of mathematical if not exactly mechanical genius. Alan Turing had worked it out in a flash of jaw-dropping insight. The bombes operated on the principle he had discovered, that if one knew or could reliably guess at the plain text that corresponded to several letters of an intercepted Enigma message, a unique pattern was formed in the way those enciphered and unenciphered letters were interrelated from one position to the next in the message; moreover, those patterns had distinct mathematical properties which were unique—or at least nearly unique— for every different setting of the Enigma, and these could be tested systematically to find the setting that had generated them. The bombes used an electric motor to drive the rotors of a battery of interconnected Enigmas through every possible position until the looked-for pattern was electrically detected. Each "hit" could then be tested by setting up an Enigma replica with the setting recovered by the bombe and seeing if it worked to decode the rest of the message into something that looked like intelligible German.

The problem with the naval Enigma was twofold. First was getting enough reliable "cribs," those bits of known contents of a message. It was a chicken-and-egg dilemma. Military wireless traffic usually contained an abundance of stereotyped phrases, such as times and dates of transmission, "from" and "to" lines, common abbreviations such as BdU. The process of matching a crib to its corresponding piece of cipher text was facilitated by a peculiarity of the Enigma machine, that a letter was never enciphered by itself: an A in the plain text never became an A in the enciphered message. So a possible crib could be slid along the coded message until a place was found where there were no "crashes," spots where the same letter appeared at the same position in both plain and cipher text. But without being able to read a few messages to begin with, it was hard to guess what a good crib might be. Moreover, as would later become apparent, the German navy ran a very tight ship when it came to cipher security; wireless operators were instructed to vary common abbreviations and standard phrases, so that BdU for example might be written BDUUU or BEF.UNTERSEEBOOTE or BEFHBR.UUUBTE, or some other variation each time it was used. Finding a way into the system depended on getting a few successes to provide an initial foothold.[37]

The other problem was the navy's indicator encryption system, which was the true nightmare. Breaking one army or air force message each day with the bombe was all that was needed to read all the rest of the day's traffic on the same network. The basic setup of the machine changed only once a day, following a printed list. The individual setting used for each message was then varied by choosing a different starting position of the Enigma's three rotors. The rotors each had 26 different positions, labeled A through Z with the letters of the alphabet, so there were $26 \times 26 \times 26$ different possible start positions, 17,576 in all. The starting position for any given message was specified by a three-letter indicator, such as GHW or QYZ, and the simple system the army and air force used for encrypting these indicators at the start of each message allowed the Bletchley code breakers to read them off as easily as their intended recipient, once they had recovered the basic setting of the day using the bombes. The navy's external code tables and more complex system for enciphering the message indicators, however, meant that breaking one message got the code breakers no further than that one message. A bombe run took hours to crack a single message; tackling every message individually was impossible.

With the code tables from *U-110* in hand the whole problem split wide open. A flood of cribs for future work was provided along with a regular trove of decoded orders to U-boats and sighting reports sent from the submarines. With traffic in hand for May and June, Turing was able to develop a way to reconstruct future months' code tables even without having any more lifted copies. Just to make sure, the Admiralty cautiously agreed to one more "pinch," and on June 25 another German weather trawler, the *Lauenburg*, was captured in the Norwegian Sea about 900 miles north of Scapa Flow with the July list of daily settings aboard. (When the captain of the British destroyer *Tartar* explained to his crew that he wanted them to fire on the trawler but not hit her, the chief gunnery mate replied, "Christ, that should be easy.")[38]

The code breakers soon discovered that some hand cipher systems used by the German navy, mainly to communicate with small ships in the Baltic, frequently carried weather reports and mine warnings that were repeated verbatim in Enigma messages sent to larger ships and U-boats. This became another fertile source of cribs for breaking future traffic; on several occasions the British were able to speed the process by laying mines with the deliberate purpose of generating a German message. By July and August 1941 information from naval Enigma signals more than once was available in time to divert a convoy around a waiting wolf pack. Sinkings by U-boats in August dropped dramatically to 80,000 tons.

CHURCHILL'S ENTHUSIASM for scientific invention was arguably exceeded only by his enthusiasm for cloak-and-dagger intrigue; that, too, caused its share of disasters throughout the war. Still, it meant that the prime minister needed no selling on the importance of the Enigma decrypts. In an initial burst of utterly unrealistic excitement, he demanded to have a copy of *every* decoded Enigma message delivered to him daily, in a special dispatch box. Later he settled for receiving a selection of the most important messages, but still insisted on seeing the actual texts, not summaries. He constantly bombarded the chiefs of staff and theater commanders with cables and memoranda calling their attention to what he considered significant bits of the decoded messages.

Churchill also took an early opportunity to show his support for the code breakers. On September 6, 1941, the prime minister paid a surprise visit to Bletchley Park. Standing on a tree stump by the lake on the grounds

of the park, he addressed the assembled staff. "You all look very . . . innocent," he jokingly began.[39]

The work was becoming much more technical as well as much more of
an assembly line than anything that GC&CS's old-school donnish linguists
had ever known, and the breakthrough on the naval Enigma now brought
the clash of cultures to a head. Bletchley's director, Alastair Denniston, was
still struggling to run things the old way; he was conscientious but simply
overwhelmed by the logistical problems of recruiting, hiring, and housing
hundreds of new staff. He reacted touchily to complaints from the naval
Enigma group throughout the summer of 1941 that they were desperately
short of manpower to maintain the bombes, carry out IBM punch card
operations that were an essential part of the job, test the bombe results on
the replica Enigmas, and even get their results typed up.

Simply, Bletchley was unprepared for success. Having achieved the near-
miraculous in cracking the naval Enigma problem, they were now unable
to exploit it to the fullest. In August the head of GC&CS's naval section,
Frank Birch, wrote to Denniston pointing out that the shortage of typists
alone was causing their output to queue up for days. Administrators were
wasting an inordinate amount of time on recruiting, but it was proving
almost impossible to get young women in particular to work at Bletchley
given the low salaries, lack of recreational facilities, inadequate meals, and
poor housing, notably the primitive plumbing that was standard in private
billets in the area—"no baths at all and the W.C. at the bottom of the garden."[40] All this did was to elicit a testy dismissal from Denniston of Birch's
"somewhat destructive memorandum," as he termed it. "What does Birch
suggest," Denniston continued, "that we should move to Harrogate or some
such place. . . . There are worse places in the country where there is not even
a cinema. There is certainly one good cinema in Bletchley."[41]

Meanwhile Dilly Knox was furiously resisting any attempts at mass-
producing the process of reading enemy messages, fighting a rearguard
action from his embattled ivory tower to preserve what he thought was his
sole right to keep control of his discoveries, rather than collaborating as part
of a team. "As a scholar, for of all Bletchley I am by birth breeding education
profession + general recognition almost the foremost scholar," he crazily
wrote Denniston, "to concede your monstrous theory of collecting material
for others is impossible . . . had the inventor no right to the development and
production of his discourses, we should still be in the Dark Ages. . . . There
are occasions when disobedience is a primary duty."[42]

Something had to be shaken up. In October several of the key Bletchley mathematicians working on the Enigma decided on an audacious move. For reasons that he said he could not remember, Stuart Milner-Barry, an international chess master who had been recruited to Bletchley in early 1940 and was now deputy head of Hut 6, which was responsible for army and air force Enigma, was chosen by the others to carry a message to the prime minister himself. It was Trafalgar Day, October 21, 1941, the 136th anniversary of Nelson's victory in 1805 over the French fleet. "What I do recall," Milner-Barry said, "is arriving at Euston Station, hailing a taxi, and with a sense of total incredulity (can this really be happening?) inviting the driver to take me to 10 Downing Street."[43]

The taxi driver did not bat an eye, and arriving at the prime minister's residence Milner-Barry boldly marched in and announced he had come from a secret war station and needed to see the prime minister immediately on a matter of national importance. He was, unsurprisingly, told that was impossible. Milner-Barry countered that he could not possibly leave the letter he was carrying with anyone but the prime minister himself, given its sensitive nature. Finally, Churchill's principal private secretary, Brigadier George Harvie-Watt, appeared; Milner-Barry was able to establish that he wasn't a raving lunatic by referring to the prime minister's recent visit to Bletchley, which Harvie-Watt knew of; and the secretary promised to see that the letter would be handed directly to the prime minister.

The code breakers emphasized in their letter that crucial work was "being held up, and in some cases not being done at all" due to the manpower shortages; the recovery of the naval Enigma keys was being delayed at least twelve hours each day; promises made back in July that the work of testing the bombe solutions would be turned over to a contingent of Wrens—members of the Women's Royal Naval Service—had come to nothing. The next day the code breakers' letter landed on the desk of General H. L. Ismay, the prime minister's chief military aide, with one of Churchill's famous red ACTION THIS DAY labels affixed. Beneath the label Churchill had scrawled a succinct instruction: "Make sure they have all they want on extreme priority and report to me that this has been done."

Within a few months Denniston was out; orders for more bombes were issued; the Ministry of Labour was ordered to meet with the head of the British Secret Service and arrange whatever manpower Bletchley needed; the army, navy, and air force were told to make additional servicemen and servicewomen available. Bletchley's staff would reach about 1,500 by the end of the year, and from the naval Enigma group in Hut 8 a steady stream of decrypts flowed to the Admiralty's Operational Intelligence Centre, where U-boat positions were updated with pins on a huge wall chart.

The combined effect of Bletchley's breakthrough in reading U-boat messages and the growing efficiency of British antisubmarine patrols as they began to implement the recommendations of Blackett's team delivered a sharp check to Dönitz's offensive in the second half of 1941: the "Happy Time" was over. Monthly sinkings by U-boats, which had been averaging 250,000 tons in the first half of the year, dwindled to half that figure as convoys were safely rerouted around the lurking wolf packs, while the number of U-boats sunk doubled.

Puzzling over his abrupt change of fortune, in particular the failure of his boats to intercept one expected convoy after another, Dönitz confided his frustrations and suspicions in his war diary entry of November 16, 1941. "Coincidence does not fall on the same side every time," he wrote. Yet nothing really seemed to explain it. It was possible the British had a new method of radio direction finding that was precisely locating the U-boats' positions at sea; there were rumors about new kinds of radar; there was always the possibility of spies or treason. All seemed unlikely. The one thing the admiral was certain of was that it was impossible for anyone to have broken the Enigma cipher, given the sheer number of mathematical permutations and the safeguards used to encrypt the setting changes. "This possibility is continuously checked by the Naval War Staff," Dönitz noted, "and regarded as out of the question."[44]

FOLLOWING GERMANY'S ATTACK on Russia on June 22, 1941, Hitler had again sought to buy time in the Atlantic. At a conference with Admiral Raeder the day before, he told his naval commander-in-chief that he desired "absolutely to avoid any possibility of incidents with the U.S.A. until the development of Operation Barbarossa"—the invasion of the Soviet Union—"becomes clearer." A month later he reiterated that he wanted "to avoid having the U.S.A. declare war while the Eastern Campaign is still in progress."[45]

But incidents were taking place regardless of any restraint Hitler was inclined to show as the U.S. Navy asserted a steadily mounting presence in the Atlantic. Following his election to an unprecedented third term as president, Roosevelt had taken another huge step toward war. In March 1941 he secured congressional passage of the Lend-Lease Act, which allowed the

United States to provide whatever arms Britain needed on credit. Even more important, the act lifted the old neutrality laws' prohibition on transporting armaments on U.S. merchant ships. American cargo vessels now were plying the dangerous waters of the Atlantic carrying tanks, planes, trucks, fuel, and steel to keep the British war machine running.

In July, U.S. Marines landed in Iceland to take over bases that the British had manned since Hitler's invasion of Denmark, which owned the island. Admiral Harold Stark, the chief of naval operations, drafted the proposed order and sent it to the White House for approval. He wrote in a covering note to Harry Hopkins: "I realize that this is practically an act of war."[46] U.S. Navy warships began convoying merchant ships to Reykjavik; though nominally independent of the interlocking Canadian and British convoy system, the American convoys were open to any neutral vessels that wished to join them. To bolster the force available for escort duty, the U.S. Navy shifted from the Pacific three battleships, an aircraft carrier, four light cruisers, and two destroyer squadrons. American warships were instructed not to attack U-boats they sighted but they could pass on the information to the British—and could defend themselves if attacked.

On September 4, 1941, *U-652* was doggedly pursued by the U.S. destroyer *Greer* for three hours. A British Hudson bomber from Coastal Command dropped three depth charges on the U-boat, which responded by firing two torpedoes at the *Greer*, which responded in turn with a depth charge attack of her own. No damage was done on either side, but it gave FDR an opportunity to commit America further. Playing down the *Greer*'s initial pursuit of the U-boat, he denounced the German action in a fireside chat a week later as "piracy legally and morally" and offered a stark warning: "When you see a rattlesnake poised to strike, you do not wait until he has struck before you crush him." Two days later he officially ordered the U.S. Navy to begin escorting Canadian convoys from Halifax as far as the Iceland meridian—a plan already secretly worked out by the British and American naval staffs— and to shoot on sight any German U-boats encountered along the way. The move immediately freed up forty British destroyers and corvettes. Far more important, it was another giant step toward intervention in the war by the United States.[47]

British government exasperation over the failure of America to formally enter the war was mounting, but Churchill told the War Cabinet that American opinion under Roosevelt's leadership had moved far faster than anyone could have expected and it was important to understand "the peculiarities of the American Constitution," which reserved to Congress the power to declare war: it was unwise for a president of the United States to get too far ahead of public feeling. London and Washington were still heaving a collective sigh of relief over the bare one-vote margin by which the U.S. House of Representatives in August passed an eighteen-month extension to the one-year army draft approved the year before. The next month the isolationist bloc in the Senate almost succeeded in scuttling an amendment to the Neutrality Act that the president sent up following his speech, to permit the arming of American merchant ships and the convoying of shipping all the way to British ports. Privately, though, Churchill shared the frustrations. "The American Constitution was designed by the Founding Fathers to keep the United States free of European entanglements," he told one of his private secretaries that fall, "and by God it has stood the test of time."[48]

But the plain fact was America already *was* at war in the Atlantic, declared or not. On October 20 the U.S. Navy destroyer *Reuben James* was torpedoed by a U-boat while escorting a convoy 600 miles west of Ireland. The ship literally broke in two, the bow section plunging under the sea instantly, the stern half five minutes later. Of the 160 men aboard, 45 were rescued out of the freezing, oil-slicked waters.

The folksinger Woody Guthrie, like other American communists, had been volubly denouncing the struggle between Britain and Germany as an "Imperialist War" that America should stay out of, adding a strident antiwar voice on the left to the anti-Semitic-tinged isolationism of Charles Lindbergh's America First movement, the *Chicago Tribune*, and others on the American right. In the first months of 1941, while America First members marched in front of the British embassy in Washington protesting Lend-Lease with placards that read BENEDICT ARNOLD HELPED ENGLAND TOO, Guthrie and the Almanac Singers were belting out protest songs denouncing Roosevelt and the war, too:

> Now the guns of Europe roar
> As they have so oft before
> And the warlords play their same old game again.
> While they butcher and they kill
> Uncle Sammy foots the bill,
> With his own dear children standing in the rain.[49]

But with Germany's attack on Russia in June, the Almanac Singers had literally changed their tune. Now the sinking of the *Reuben James* gave them the opportunity to turn their talents for agitprop to a flag-waving, full-throated, patriotic appeal for America to get into the fight. Set to the tune of the old Carter Family standard "Wildwood Flower," their song about brave Americans going down in the cold icy waters of the Atlantic was soon heard everywhere. Pete Seeger was probably the most talented member of the group, and he came up with the chorus to "Reuben James" that everyone remembered:

> *Tell me, what were their names?*
> *Tell me, what were their names?*
> *Did you have a friend on the good Reuben James?*[50]

The Irish-born inventor John P. Holland in one of his early submarines

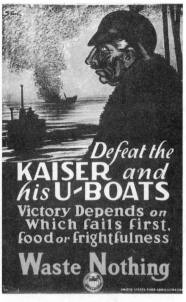

An American propaganda poster from the First World War highlights the U-boat threat

A British officer boards a German submarine as it arrives in Harwich to surrender, November 1918

The "Prof," Frederick A. Lindemann (left), with Winston Churchill

Patrick Blackett

Sir Henry Tizard, who headed the British scientific
air defense committee and repeatedly clashed with
Lindemann over science and war policy

E. J. Williams and Cecil Gordon, who along with Blackett probably contributed the
most to operational research against the U-boats

The zoologist Solly Zuckerman, at Tobruk in 1943: his Tots and Quots club and their book, *Science in War,* were instrumental in mobilizing scientific manpower for the war.

HMCS *Sackville,* one of the "cheap and nasty" flower-class escort corvettes

A convoy of merchant ships crosses the Atlantic, 1942

A pilot of a Sunderland flying boat scans the sea on a U-boat patrol over the Atlantic

Vannevar Bush, the MIT engineer who galvanized American defense research, in a characteristic pose

U.S. Navy captain Wilder D. Baker, who brought American scientists into the anti-U-boat fight (shown here in 1944 receiving the Navy Cross)

MIT's Philip Morse, the physicist who directed ASWORG, and his deputy William
Shockley of Bell Labs

Air Marshal John Slessor, who clashed with
Blackett over RAF bombing policy

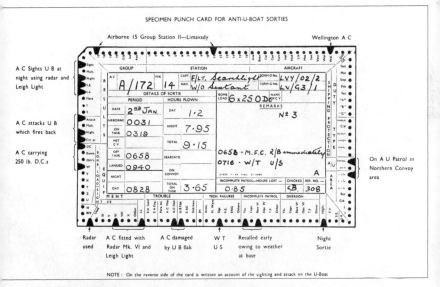

A cartoon adorning the back cover of an ASWORG report

SPECIMEN PUNCH CARD FOR ANTI-U-BOAT SORTIES

Punch card used by Coastal Command ORS to analyze attacks on U-boats: these cards stored considerably more information than those used by IBM machines, but had to be sorted by hand (using knitting needles) to select cards with specified criteria.

One of the U.S. Navy's electromechanical "bombes" used to recover the daily settings of the U-boat Enigma codes

U-118 under depth charge and machine gun attack by aircraft from the U.S. escort carrier *Bogue,* June 12, 1943

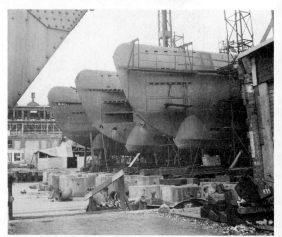

U-boats under construction in Hamburg at the war's end

8

Baker's Dozen

THE JAPANESE SPARED FDR the need to exercise his powers of persuasion any further on an isolationist and war-wary public. "We are all in the same boat now," the president told Churchill when the British prime minister reached him by telephone the evening of the Japanese attack on Pearl Harbor on December 7, 1941.

A formal declaration of war against Germany still depended upon what Hitler would do; there were four tense days in Washington waiting to see what would happen. On December 11, Hitler addressed the Reichstag with a vituperative personal attack on the American president that he had been obviously holding in for months. He called Roosevelt a hypocrite, a Freemason, a rich financial speculator who surrounded himself with Jews, a "man who, while our soldiers are fighting in snow and ice, very tactfully likes to make his chats from the fireside." He ended to thundering applause with a declaration of war against the United States. Roosevelt sent a message to Congress the same day asking for a declaration that a state of war with Germany existed; it passed unanimously.

In London, Churchill felt nothing but unconcealed relief. "So we had won after all!" was his first thought. "No American will think it wrong of me," he later wrote, "if I proclaim that to have the United States at our side was to me the greatest joy":

Many disasters, immeasurable cost and tribulation lay ahead, but there was no more doubt about the end. Silly people—and there were many, not only in enemy countries—might discount the force of the United States. Some said they were soft, others that they would never be united. They would fool around at a distance. They would never come to grips. They would never stand blood-letting. Their democracy and system of recurrent elections would paralyze their war effort. They would be just a vague blur on the horizon to friend or foe. Now we should see the weakness of this numerous but remote, wealthy, and talkative people. But I had studied the American Civil War, fought out to the last desperate inch. . . . I thought of a remark which Edward Grey had made to me more than thirty years before—that the United States is like "a gigantic boiler. Once the fire is lighted under it there is no limit to the power it can generate."[1]

The day after war began between Germany and the United States, Hitler ordered Raeder to send six of the large Type IX U-boats to the east coast of America "as quickly as possible."[2] The news that America was officially in the war was as welcome to Dönitz as it was to Churchill. The British successes over the last several months in evading and destroying his submarines—the product both of the British scientific breakthroughs in operational research and cryptanalysis, and of a growing organizational ability to put those discoveries promptly to use—had severely checked the U-boat offensive in the Atlantic. For the second half of 1941 merchant shipping losses to U-boats were averaging under 125,000 tons a month. More than that, the individual effectiveness of U-boats had dropped dramatically. Each boat at sea had sunk an average of four ships a month during the "Happy Time"; that figure was now down to less than one per month.[3]

On January 13, 1942, Dönitz's advance force of six U-boats reached their positions and from BdU headquarters received a transmitted code word, *Paukenschlag*, "drumbeat," or maybe more precisely "timpani crash." It was the signal to begin the war with the United States. The result was mayhem. Only five of the Type IX boats had successfully completed the voyage across the Atlantic but over the next two weeks they sank thirty-five ships totaling more than 200,000 tons. Tankers billowing thick columns of black smoke could be seen each morning in plain view off the shores of Cape Cod, New Jersey, the Virginia Capes, and Cape Hatteras.

In February a second wave of U-boats arrived to take over from the first five, which were by then running low on fuel, torpedoes, and supplies; the new arrivals included some of the standard medium-range Type VII Atlantic boats with extra fuel tanks crammed in every available space to extend their cruising distance. They, too, found a U-boat paradise awaiting them. It was, the U-boat captains said, the "Second Happy Time." Freighters crawled along the coast singly, often with lights on, or silhouetted against the blazing lights of coastal cities that had yet to order blackouts. In his war diary Dönitz described conditions on the American coast as "almost of a peacetime standard," with "single-ship traffic, clumsy handling of ships, few and unpracticed sea and air patrols and defenses." The pickings were so rich that his commanders could coolly let freighters in ballast pass by and wait for a more lucrative tanker or laden freighter to appear, as it inevitably would. The first of the U-boat captains to return reported that there were so many targets available he could not possibly engage them all: "At times up to ten ships were in sight sailing with lights on peacetime courses."[4] The rampage extended into the Gulf of Mexico and the Caribbean. In March sinkings reached a half million tons. Dönitz was promoted to full admiral.

The slaughter would have been far worse if Hitler, who never could quite grasp naval strategy, had not at that moment persuaded himself that the Allies were planning a move against Norway. By spring of 1942 the total number of U-boats in the German fleet was approaching 300 thanks to an ever-accelerating tempo of new construction; 170 or so were undergoing training and trials, but that left more than 100 available for operations. About 50 boats were now at sea at any given time. In vain Dönitz pleaded to throw them all into the western Atlantic. But with his eyes on Norway, Hitler overrode Dönitz, holding back a sizable force to repulse the Baltic thrust that never came. Still, even with one hand tied behind his back by his Führer, Dönitz was delivering a terrible pounding to Britain's new ally. One day that March, the MIT physicist Philip Morse was on a ferry from Delaware to Newport News, Virginia, when the ship passed a tanker limping to port with a ten-foot gash torn through her hull near the bow, obviously the victim of a German torpedo attack. Morse wondered who was analyzing the U-boat threat for the U.S. Navy.

AS MORSE HIMSELF would learn a few weeks later, the answer was no one, though it had not been for want of admirable intentions. In 1940, at the same time Vannevar Bush was securing the president's blessing for his plan for the NDRC, Frank Jewett was enlisting the support of the secretary of the navy for the creation of a National Academy of Sciences committee to advise the navy on scientific matters. In December 1940 the academy's Naval Advisory

Committee organized a subcommittee to look specifically at problems of antisubmarine warfare; Jewett chose as its head Edwin H. Colpitts, a former vice president of Bell Labs. The Colpitts Committee dutifully spent two months touring naval facilities and examining the navy's existing technical programs. In its final report, the committee recommended that the government institute a broad-based scientific program on oceanography and underwater acoustics, explore new devices based on centimeter-wave radar and magnetic locators that had been neglected to date due to a shortage of funds, and improve training and maintenance so that sonar equipment already in the field could be properly employed, which it frequently was not.

One of the Colpitts Committee's additional recommendations to the navy was the formation of a committee to investigate the entire problem of submarine detection. The fix was in on this: even before the committee reported, Bush had written Jewett that the antisubmarine problem was "absolutely the kind of thing on which NDRC ought to take off its coat and get busy," and Jewett and Colpitts fully intended that NDRC would get the job. In April 1941 Jewett arranged for a formal request from the navy to NDRC to implement the Colpitts Committee's recommendation. Bush and Jewett chose John T. Tate, dean of the University of Minnesota, professor of physics, and editor of the prestigious *Physical Review,* to head the new Section C-4 of NDRC that was created to take on the navy assignment. Columbia University was asked to serve as the contracting agent, and by the summer of 1941 about $1.5 million in contracts was flying out the door to researchers in various universities and industries, including General Electric, RCA, Harvard, the Woods Hole Oceanographic Institution, and the University of California.[5]

Following up on a suggestion by Tizard during his visit to the United States the previous year, Tate and Louis B. Slichter, an MIT geophysicist who had worked on submarine detection in the First World War, left for a visit to England on April 7, 1942, to continue the technical liaison between American scientists and the Admiralty. They returned a month later with a bagful of notes and technical documents: drawings of hydrophones, plans for an "attack trainer" to teach escort ship crews how to use asdic, a booklet on underwater vibration patterns of propeller blades.[6]

It was all important information, but like the focus of Section C-4 itself it was heavy on the technical details of apparatus and acoustical phenomena with little about actual antisubmarine tactics and strategy. Most of the NDRC's contracts involved basic scientific research in oceanography, underwater sound transmission, and magnetic and optical principles of detection. Phil Morse was brought into the work of Section C-4 in the summer of 1941 via his background in acoustics, and it did not take long for him to chafe at the narrow, technical confines of the tasks being portioned out to the scientists by their navy patrons.

Morse was then thirty-eight years old. A dapper midwesterner who sported a Clark Gable mustache and pin-striped suits, he was always caught in photographs with an ever-so-slight look of wry amusement in his eyes. He came from a family of "three generations of technical men, builders, and planners," in his words; his ancestors included a railroad surveyor and a telephone engineer, a builder and architect, and several committed abolitionists and social activists. Morse had found MIT a perfect home. He had been on the physics faculty for a decade, arriving in 1931 as part of MIT president Karl Compton's ambitious plan for the United States to begin training its own scientists, rather than relying on English and German universities as it always had before. Compton was especially eager to convince industry that physicists, as scientific generalists, could help them solve any technological problem. It was an approach that was beginning to pay off; under Compton's presidency the physics department expanded rapidly. When Morse arrived there were about a dozen physics graduate students; six years later there were sixty.

Morse had been Compton's student at Princeton, which he had chosen for graduate school knowing little more about the place than the fact that Princeton had offered him a $700 scholarship, versus Harvard's offer of $450. Under Compton he had done an experimental study involving electrical discharges in gases. He spent months blowing glass and building electrical devices and assembling a mercury vapor vacuum pump. Earlier, as an undergraduate at the Case engineering school in Cleveland, he had helped pay his way when his family ran short of money by taking a year off and working as a partner in a radio store where the owners assembled much of the equipment themselves. But his experiences in grad school convinced him that he was not a natural experimentalist, and he found himself increasingly drawn to the theoretical side of physics.[7]

At MIT it dawned on him that he wasn't a theoretician either, at least not the kind who would ever win a Nobel Prize. "It was clear to me," he later wrote, "that I was no Einstein." ("This realization," he added, "came slowly enough to cushion disappointment.") But Morse was becoming just the kind of generalist that Compton thought made physicists so useful. Morse recognized that his own skill was "not the sort of deep-thinking ability that wins prizes and fame," he admitted, but rather the ability to look across many areas, find the connections, the promising leads, and see how the insights in one field could be applied to another. He enjoyed teaching, writing textbooks, guiding graduate students, working on an interesting problem for a few weeks, then moving on to something new. "Breadth rather than depth was best for me," he concluded. Morse's interest in acoustics had come about that way, by noticing that the mathematical techniques of quantum mechanics could address interesting questions in sound propagation.[8]

His years at MIT in the 1930s, he later thought, were "a happy untroubled time of a sort that has never since returned." He was earning $7,500 a year by 1939 from his salary as a full professor plus consulting fees and textbook royalties; he was settled in a happy marriage, with a home in the comfortable and affluent Boston suburb of Belmont that his in-laws helped the couple buy, and his wife had just given birth to their first child, a daughter; he had time to indulge his restless and eclectic hobbies—mountain climbing, wood turning, a voracious appetite for reading, usually five books a week, history, archaeology, and biography mostly.

Much like Blackett, though, he was at least at times struck by the disconcerting gap between scientific progress and the hardships suffered in society at large. Struggling in the Depression back in Cleveland, his mother took in boarders; his father was out of work and ended up taking a job with the WPA, planning a building that never got built. A broken and discouraged man, he aged rapidly and died in 1939. "At MIT I was helping to discover new things," Morse wrote of one of his moments of self-doubt, "while the country was not able to use the ample resources and technology it already had to feed and house its people." He had been apolitical like many of his fellow scientists, believing there was little he could do to influence public events. He was also a pacifist. Like many of his British counterparts, he was shaken out of both beliefs by the rise of Nazi Germany. He reached two conclusions. One was that he would contribute whatever he could to the war effort. His corollary was that if scientists were to play a role, it was essential that "scientific work for the war effort should not be entirely controlled by the military and that scientists must have a part in deciding, at the highest levels, what direction their work would take."[9]

Morse spent the summer of 1941 shuttling between Cambridge, Cape Cod, and various navy laboratories developing a noise-making device that could be towed behind a ship to counter acoustically triggered mines the

Germans were reportedly developing. His design, two hollow parallel steel pipes that generated a deafening cavitation noise—the sound produced by water pulling away then smacking back against an underwater surface— won the record for the loudest sound per weight of any of the devices tested. But by the end of 1941 "some of the glamour had left the project for me," Morse found:

> True, we had done what we had been asked to do, and had done it quickly. We were proud to be able to show what scientists could do. Most of us liked to build equipment and were enjoying the chance to put to use new knowledge and techniques amassed during the past fifteen years. However, some of us wondered whether the only thing trained scientists were good for in a war was to do the measurements and design work thought up for them by the supply departments of the armed services. Having become acquainted with many of the officers in charge of projects, I entertained a faint doubt as to whether these officers were always asking us to do the right things.[10]

Morse shared his frustrations with Tate, an old friend. Tate listened sympathetically but said little. Tate's tight lips were themselves a symptom of the deeper problem, Morse realized. The navy fundamentally distrusted sharing its secrets with civilians, and Tate was having to walk an extremely fine line between trying to expand the role of the scientists and not alarming naval officers that he was poking into areas where he had no authority; that meant he had to keep his mouth shut far more often than he otherwise would have liked to.

TATE WAS in fact quietly thinking about Morse's complaint. A few months later, in mid-March 1942—not long after Morse had spotted the crippled tanker from the Newport News ferry—Tate asked him to pay a call on Captain Wilder D. Baker at the navy's First Naval District headquarters in Boston. Baker's office was in the office building at North Station, the terminal of the Boston & Maine Railroad across the Charles River from the Charlestown Navy Yard. Morse was favorably impressed with the navy captain the moment he walked in the door: "steel-gray eyes, gray hair, a look of decisiveness."

Baker was in many ways a completely conventional and unremarkable career navy man. He had a reputation as a demanding, stern, but fair and direct commander; when he promised to do something, he did it. His father had been a newspaperman in Topeka, Kansas; the family later moved to Bay City, Michigan, and on graduation from high school he had applied for admission to the Naval Academy. He graduated in the middle of his class at Annapolis: his entry in the 1914 *Lucky Bag,* the academy yearbook, has a joking allusion to his being more interested in sleeping than studying, as well as some good-natured ribbing about his inability to keep step with the drum on parade and his popularity with girls ("yes, the girls *fight* for him"). His other off-duty interests were almost stereotypically conventional. He was an avidly enthusiastic—though, his son recalled, "terrible"—golfer; he went shooting and fishing. His career had followed the standard progression of promotion through increasingly responsible sea commands. As a lieutenant in the 1920s he commanded the submarines *S-11* and *S-13;* he was a lieutenant commander on the battleship *Wyoming* in the 1930s; and on October 1, 1941, he was promoted to captain and given the command of Destroyer Squadron 31 in the Atlantic. That last sea command had given him at least one very direct encounter with U-boat warfare: the *Reuben James* was one of the nine ships in his squadron when she was torpedoed later that month.

There were a few byways in his career, though, that hinted of a man with broader experiences and outlooks. In 1934 and 1935 he was assigned as a naval attaché in London, his real job being to observe the Italian-Ethiopian war. And in 1938 he was sent by the navy to New Haven, Connecticut, to head naval science instruction for Yale's ROTC program. "He was only a commander, just coming off sea duty, and it was a challenge for him" to be thrust into such an academic environment, his son said. "But he treasured those two years and the friendships he formed there more than any in his life." He became close friends with a number of Yale's leading faculty and administrators.[11]

On February 7, 1942, Baker took over the command of the newly established Anti-Submarine Warfare Unit of the Atlantic Fleet.[12] It was, remarkably, the U.S. Navy's first attempt at coordinating and supervising antisubmarine training and tactical doctrine. Baker's first step was to spend about a month in England seeing what he could learn from the British experience fighting the U-boats. One of the people he met there was Patrick Blackett, who two months earlier had become chief advisor on operational research to the Board of Admiralty.

BLACKETT'S MOVE to the Admiralty was a belated recognition by the British navy of the revolution in antisubmarine warfare that Blackett and his group had already been effecting at Coastal Command. It had come about almost completely as a result of his "Scientists at the Operational Level" paper, which he had presented in September 1941 to an Admiralty scientific advisory panel chaired by the physicist E. V. Appleton. The panel afterward recommended that Blackett be brought in to do for the navy as a whole what he had done already for Coastal Command. At last fully awakened to the importance of operational research, the Admiralty on December 10, 1941, created the new position and Blackett moved into his new office in Room 74 of the Old Admiralty Building.

As usual, it was a bureaucratic improvisation. Blackett was officially an adviser to the board while also reporting to the controller of the Admiralty and the vice chief of the Naval Staff; the other scientists working with him were attached to other parts of the navy, such as the gunnery, antisubmarine warfare, and signals divisions; the entire operations research group was meanwhile, and simultaneously, a subdepartment of the navy's Scientific Research and Experiment Department. The navy was not willing to allow him more than three senior scientists plus a few junior scientists, a much smaller group than he had at Coastal Command, and an almost comically lengthy exchange of minutes went on for months wrangling over the details of his salary and concerns over the bureaucratic precedent that would be set were he to receive the same pay as a director of a much larger research establishment. It was finally agreed that he would receive a "special allowance" of £200 in addition to his salary of £1,000, and that the University of Manchester would make up an additional £500 of his pay.[13]

Blackett's arguments and his "Scientists at the Operational Level" paper impressed Baker as they had the British Admiralty. Baker returned to Boston in March 1942 with a copy of Blackett's now increasingly famous manifesto, which he quoted from at length in a memorandum proposing that the U.S. Navy follow the British example and assemble a team of eminent scientists to undertake a thorough examination of operational tactics in the antisubmarine war. Writing to Rear Admiral J. A. Furer, the U.S. Navy's coordinator of research and development, on March 16, Baker argued that while any experienced officer could assess the results of a single attack, developing consistently effective methods to counter the U-boats required building up a body of accurate records and statistics—and having men who were trained to analyze those results and sift them for patterns and trends.

"I do not suppose that any really good naval officer is qualified for mathematical or scientific analysis (he would not be a good naval officer if he were)," Baker wrote, "and that therefore the man selected to organize and run this group had better be left to do only that."[14]

There was little doubt that Baker had been completely sold on Blackett's vision of what operational research was, and what it could do for the navy. In his memorandum to Furer—apparently written just before his first meeting with Morse—he almost directly echoed Blackett's words:

> The analytical section should be outstanding men of reputation with broad vision and receptive minds, able quickly to comprehend the needs and problems with which we are confronted, and experienced in utilizing the abstract as well as the material tools of science in solving such problems.[15]

At their first meeting, Baker and Morse warily sized each other up. They began by talking in generalities about the work each was doing. Baker apparently decided that Morse was all right, because after a few minutes he abruptly shifted gears and explained what he was really after. It was, Morse later remembered, by far the longest speech he would ever hear the normally terse navy man deliver. It did not take Morse long to jump at the offer. It was also apparent to him that Baker was taking a huge personal risk. Though for convenience the scientists would be paid through the NDRC and Columbia University under an existing navy contract, operationally they would be part of Baker's unit, which was a distinct departure for the navy. As Morse recalled:

> Baker was willing to give me a chance to show what a civilian task force could do. To let nonmilitary persons participate in even minor operational decisions was, of course, heretical to many officers, especially those in the Navy, with their tradition that the commander of the ship or the fleet was the absolute master. But Baker had seen enough, in Britain, of the urgency and complexity of anti-submarine warfare to convince him that traditional policies were inadequate here. . . . He never said so explicitly, but it was soon apparent to us that he had put his career on the line; if our group didn't pay off, Captain Baker would never be Admiral Baker.[16]

As in Britain, many physicists in America had already been snapped up to work on radar—principally at MIT's new Radiation Laboratory—

and on the preliminary research on the atomic bomb already under way at Columbia, Chicago, and Berkeley. "It is emphasized that it is quite difficult to obtain men of the proper qualifications who are also cleared, since such men are invariably at present at work on other defense jobs," Morse reported a few weeks later. At Tate's suggestion, Morse hired his former student William Shockley, now a researcher at Bell Labs, as his assistant supervisor. Morse also had the inspired thought of recruiting insurance actuaries, who were all mathematicians thoroughly versed in probability theory and experienced in applying it to practical questions. By July he had a staff of 13 in place: inevitably they were dubbed "Baker's Dozen." By the following year the group would grow to 44 with a makeup as eclectic as that of Blackett's Circus: 6 mathematicians, 14 actuaries, 18 physicists, 3 chemists, 2 biologists, and an architect. The sign on their door at the navy's Boston office identified them only as "Columbia Group M" (Morse said he thought the "M" might stand for Morse, but was never sure).[17] That name was just a cover, however, used for the Columbia contract and unclassified paperwork. Internally, they had now an official designation that declared their mission to the rest of the navy loud and clear. The American scientists decided to call their new discipline "operations research," slightly shortening the preferred British term "operational research," and Morse's group was dubbed the Anti-Submarine Warfare Operations Research Group, ASWORG.

THE COMMAND STRUCTURE of the U.S. Navy was a holy mess. Responsibility for antisubmarine warfare was split up among a dozen different commands; no one was in charge. Actual operational control of antisubmarine patrols along the Atlantic coast rested with the admirals who commanded the separate naval districts based at each of the major ports (Boston, New York, Philadelphia, Norfolk, Charleston, and Miami), but the ships themselves belonged to the Atlantic Fleet. The commander of the fleet was supposed to be in charge of all antisubmarine training but had no authority to issue orders to the district commandants. Overlapping the naval districts, another layer of authority had been hastily added in 1941: the "sea frontiers," which were a halfhearted attempt to establish higher-level coordination for the defense of shipping lanes on the approaches to the American coast. The Eastern Sea Frontier was based in New York, with responsibility for the coast from Maine to Florida; there was also a Gulf Sea Frontier and a Caribbean Sea Frontier. The launch of the U-boat offensive along the American coast had thrust the Eastern Sea Frontier front and center, but in 1942 it was still little more than a paper command, with an office on the fifteenth floor of the Federal Building at 90 Church Street in lower Manhattan. Though administratively equal to the Atlantic Fleet, the sea frontier commanders, like the naval district commanders, had no destroyers or other ships permanently assigned to them; they could only "request" the Atlantic Fleet to detach ships to them. Overall responsibility for developing antisubmarine doctrine and procedures rested in yet another place, the staff of the U.S. Fleet commander in Washington.

The position of commander-in-chief, U.S. Fleet—CominCh—was also something new in the navy command structure. FDR had ordered the creation of the post in the aftermath of Pearl Harbor; for the first time, it gave direct operational authority over the entire navy to a single top admiral, based in Washington. While a much needed step in theory, in practice the immediate effect was to sow even more chaos and confusion in the lines of authority throughout the navy, the antisubmarine campaign in particular.

Some but not all of that chaos stemmed from the personality of the man chosen for the job. Admiral Ernest J. King was brilliant, capable, and confident. He was also bullheaded, ruthless, and vindictive. In 1939 everyone in the navy, King included, had been certain his career was over. Passed over for the position of chief of naval operations, the service chief of the navy, he had been shunted off to the General Board, the place where over-the-hill admirals were put out to pasture, whiling away their few years before retirement writing studies that sat unread on the shelves of the secretary of the navy's office.

At age sixty, King had made almost nothing but enemies in his four-decade career in the service. He bad-mouthed rivals, contemptuously dismissed subordinates, bristled at even mild advice from superiors; off hours he drank like a fish and remorselessly chased fellow officers' wives. "His appeal to women was most unusual," remarked one wondering officer.[18] The appeal was apparently that he could be charming on social occasions, was a good dancer and an intelligent and interesting conversationalist, and did not hesitate to proposition any attractive woman who came along. His wife remained installed in a home in Annapolis he had purchased shortly after they were married in 1905 and raised the couple's six daughters and a son while King followed his naval career, and other pursuits.

There was no doubt, though, that King got things done. He was always more determined, more energetic, more demanding than anyone around him. His sheer stamina was legendary. He was no intellectual but none-theless possessed a formidable natural intellect, and the drive to put it to use when required. At age forty-seven he had earned his wings so that he could command an aircraft carrier. After a moment of despair upon being assigned to the General Board, he astonished everyone by taking the job seriously, throwing himself into a series of complex studies, putting in improbable hours at his desk in Washington. "They're not done with me yet," he told an acquaintance he ran into in a Navy Department corridor.[19]

In December 1940 Secretary of the Navy Frank Knox was looking for a commander to take charge of Atlantic escort operations and decided to bring King off of the sidelines—for one last tour. King immediately put his ships on a war footing and made it clear they were going to take the offensive in the Atlantic in the as yet undeclared war with the Nazi U-boats. In the aftermath of the destroyer *Greer*'s depth charge battle with a U-boat in Sep-tember 1941, he called in the ship's commander to reassure him. "As long as I command the Atlantic Fleet," King told him, "no one is going to nail your tail to the mast because you defended yourself."[20]

The coming of the war rescued King's career. "When they run into trouble, they send for the sons of bitches," he reportedly commented on his change of fortune after being appointed CominCh. The most famous obser-vation about King was one attributed to his daughter: "My father is the most even-tempered man in the navy. He is always in a rage." Roosevelt had heard the other joke that had been making the rounds about King for years, that he "shaved with a blow torch," and the president sent him a letter quoting the line and saying how glad he was to have "the toughest man in the navy" in charge. FDR added: "P.S. I am trying to verify another rumor—that you cut your toenails with a torpedo net cutter."[21] There was no doubt that the navy, and the country, needed someone with King's aggressive spirit to put some confidence back into the navy in the wake of Pearl Harbor.

But in the rapidly deteriorating war with the U-boats along the coast of the United States, nearly everything King did for the first few months only muddied the waters further. The formation of Baker's ASW unit within the Atlantic Fleet had been a small step toward bringing some central coordina-tion to the effort. It was undermined by King's own repeated insistence on the very traditional navy view of decentralized authority: the captain of the ship was the ultimate word. Upon taking command of the Atlantic Fleet the year before he had promulgated a famous order on the "initiative of the sub-ordinate," lambasting flag officers for issuing instructions to the individual ship's captains under their authority telling them "how" as well as "what" to do. As CominCh, King insisted at first on keeping the staff to an unrealistically small size, no more than 300 officers. When the Atlantic Fleet tried to issue its first manual for antisubmarine warfare, the fleet's staff first had to receive approval from CominCh—which refused to print enough copies to distribute to every ship. When Vice Admiral Adolphus Andrews, the commander of the Eastern Sea Frontier, issued an order to one district commandant spelling out antisubmarine patrol procedures, King "practically hit the ceiling," recalled one aide. The commander-in-chief sent a curt signal canceling Andrews's order and telling him off: "Do not presume you are on the bridge of every ship under your command."[22]

By the same token, King refused to override Andrews's misbegotten decision not to organize convoys of coastal traffic even in the face of the mounting losses to the U-boats. Andrews argued that until he had enough escorts it would be best to let the freighters and tankers continue to sail individually. Weakly escorted convoys, he insisted, would be worse than no convoys at all. In fact, even *unescorted* convoys were an effective counter to U-boats, particularly those operating singly as Dönitz's *Paukenschlag* forces were. A convoy substantially reduced the opportunity for a U-boat to find a target in the first place, and did not significantly increase the chances that more than one ship would be sunk even when a convoy was spotted.

Much has been made of King's anglophobia as an explanation for the American failure to institute convoys or otherwise benefit from British knowledge and experience during those disastrous opening months of the war for America. King did make it abundantly clear that he would be damned if the United States was going to play second fiddle to the Royal Navy and dismissed out of hand a British proposal to place the joint Anglo-American naval forces in the Atlantic under a single—British—admiral. But the far more important factor was that the U.S. Navy lacked the command structure to put any British knowledge to use even if it cared to. The shortage of surface ships and aircraft did not help, but that was not the main problem either. The main problem was that all of the various pieces of the antisubmarine campaign, from training to doctrine to operations, were parceled out among various commands, all of which had other duties as well. No one commander was actually responsible for understanding the whole picture, or doing something about it.

That was the real lesson the United States needed to learn from its more experienced ally. Coordination with air forces, or rather the lack thereof, was especially telling. The British services had been through a tug-of-war between the navy and the air force over control of naval aviation very much as the U.S. Army and the U.S. Navy had, but far more important than the decision to place RAF Coastal Command under Admiralty control in April 1941 was the fact that the headquarters for the two most important air and sea commands fighting the U-boats—Western Approaches Command and Coastal Command's No. 15 Group—were located side by side in the Liverpool command center. They closely coordinated air patrols with surface escorts and convoys. The day after Pearl Harbor the U.S. Army Air Forces I Bomber Command began sea patrols along the Atlantic coast of the United States, but it took until the end of March to work out even a preliminary agreement clarifying the command relationships and giving authority over antisubmarine air patrols to the Eastern Sea Frontier headquarters in New York. It was May before a Gulf Task Force from I Bomber Command was detached to Charleston, South Carolina, under a similar arrangement with the Gulf Sea Frontier.[23]

Worst of all was the intelligence situation. The U.S. Navy's Office of Naval Intelligence had long since become a retirement home for incompetent officers who washed out in sea commands. Its key functions were thoroughly cannibalized by other departments of the navy. The aggressive commander of the War Plans Division, Rear Admiral Richmond Kelly Turner, had wrested control of operational intelligence away from ONI even though he had no experience or knowledge of intelligence matters. (He was, though, a master intellectual bully: he had once taken a crash course in music just so he could convincingly chew out the band on a ship he commanded.)[24] The Office of Naval Communications meanwhile seized effective control of all signals intelligence and code breaking in another bureaucratic skirmish. ONI was left with compiling static assessments of the order of battle of foreign navies, collecting attaché reports that were little better than what could be had from clipping local newspapers, and carrying out often clumsy and incompetent security investigations to ferret out real or imagined spies and domestic subversives.

A small Convoy and Routing Section had been established at navy headquarters in Washington in November 1941 to organize and keep track of the transatlantic convoys, but it did not deal with intelligence about U-boat locations. It was again left largely up to the individual district and sea frontier commanders to pick up the ball and try to maintain their own map plots with their own small intelligence units.

The British Operational Intelligence Centre was almost everything the U.S. Navy's intelligence system was not. It brought together intelligence from all sources—radio direction finding from U-boat transmissions, Enigma decrypts, sightings by aircraft and surface ships, prisoner interrogation reports, and a growing file of accumulated clues about individual U-boat captains and their tactics and habits—to maintain a constantly updated picture of U-boat movements and dispositions in relation to current convoy locations. The OIC also had the authority to forward their findings directly to commanders in charge of convoy routing and antisubmarine operations.

The resident genius of the OIC's Submarine Tracking Room was a remarkable man, Rodger Winn. The ravages of polio contracted in childhood had left him twisted and gnomelike, with a terrible limp and a severely crooked back. Thirty-eight years old, he was a barrister in civilian life. His deformity would have normally made him ineligible for naval service at all but the navy arranged a direct commission for him as a commander in the Royal Navy Volunteer Reserves. Winn had exactly the kind of mind and instincts for the job, the sharp lawyer's ability to piece together often contradictory and incomplete data to form a picture of the whole, and was legendary for his almost sixth sense—or so it seemed to those not as steeped in operational intelligence as he was—to know where the U-boats were going to be next.

On April 19, 1942, Winn arrived in Washington hoping to convince his American counterparts that they needed a thorough reorganization of their intelligence system to stem the tide of disasters besetting Allied shipping in American waters. It took him several days to get in to see Rear Admiral Richard S. Edwards, King's deputy chief of staff. Winn made his case that the U.S. Navy needed to centralize its submarine intelligence just as the British had. Edwards, displaying a full measure of U.S. Navy anglophobia, retorted that the United States did not need the British to teach them anything, and moreover that if America wanted to lose ships that was her business: America had plenty of ships and could afford to lose them. Edwards also blithely insisted that it was impossible to forecast U-boat movements, thus futile to try to reroute convoys around them.

Winn, who had spent time in the United States as a student at Harvard and Yale, was well aware of what he described as the American bent for straight talking and decided to let Edwards have it. "The trouble is, Admiral," he began, "it's not only your bloody ships you're losing, a lot of them are ours. And we're not prepared to sacrifice men and ships to your bloody incompetence and obstinacy!"

Edwards was taken aback for a moment but then laughed and said, "Well, maybe you have a point there." Winn also hinted that if the U.S. Navy did set up such a centralized, and secure, intelligence operation, "we might have better information to impart if we could be sure how it would be handled." That was an allusion to the Enigma traffic—the one secret the British remained loath to completely share with their ally, given their fears about lax American security and the terrible risk of losing this source altogether if the Germans caught even a hint of its having been successfully broken. But after an "alcoholic luncheon," Winn subsequently reported, he and Edwards "parted the best of friends" and the American admiral agreed that Winn ought to be brought in to see Admiral King. Things began to happen. By the time he left Washington in May, Winn was able to report that the American U-boat tracking room was "a going concern." The operation was made part of the CominCh staff at navy headquarters and the small tracking operation at Eastern Sea Frontier headquarters in New York was disbanded, a significant reversal in King's philosophy of decentralization and a recognition that drastic changes in business as usual were going to be needed if the ravages of the U-boats were to be stemmed.[25]

THE BAD NEWS CONTINUED unabated through the spring of 1942. Even with a belated blackout of coastal cities in April and the beginnings of a more unified coordination of antisubmarine operations, the situation remained, to quote King's own assessment, "desperate." A second wave of U-boats arrived in the Caribbean around April 12 and again began picking off tankers with abandon. As a stopgap the navy established protective anchorages off Cape Lookout in North Carolina so ships could wait out the dangerous night and pass Cape Hatteras by daylight.

In mid-May convoys were finally organized from Key West to Hatteras; at once sinkings were cut from 25 percent of unescorted vessels leaving U.S. ports in May to 2 percent of those traveling in convoys. Still, the worst month was June, when 127 ships in the Atlantic theater went down, 637,000 tons in all, the grimmest toll of the war to date. The United States had lost 5 percent of its available shipping in just the first six months since it entered the war.[26] General George Marshall, the army chief of staff, normally the most reticent of men in criticizing his military colleagues, sent King a memorandum on June 19 expressing his anxiety:

> The losses by submarine off our Atlantic seaboard and in the Caribbean threaten our entire war effort. . . . We are well aware of the limited number of escort craft available, but has every conceivable improvised means been brought to bear on the situation? I am fearful that another month or two of this will so cripple our means of transport that we will be unable to bring sufficient men and planes to bear against the enemy in critical theaters.

King responded testily that he had "long been aware" of the implications of having ships sunk by U-boats and was doing everything he could.[27] He responded with an even more imperious statement to the press—his disdain for publicity and reporters was legendary—suggesting that "the volume of criticism of the conduct of the anti-submarine campaign" was hurting the war effort: "It must be obvious that we of the Navy are even more concerned than are any of the critics or any of the other citizens of the U.S. because we have the responsibility and the critics have not."[28]

All kinds of helpful suggestions poured in from members of the public. A memorandum forwarded to Captain Baker from navy headquarters dryly summarized them: "Many letters suggest we do more patrolling, convoying without giving any specific recommendation." Others offered the idea of using railroads and inland waterways to move goods, or establishing "safe lanes" for merchantmen to follow which would be constantly patrolled. Many proposed deploying small fishing boats and pleasure vessels to the hunt for U-boats ("here is always mentioned the advantage of sailing vessels as a listening post due to the lack of machinery noise").[29] The president, an enthusiastic yachtsman, unfortunately added his voice to this last idea; he ordered King to establish an armada of civilian volunteers to patrol the shores. The coast guard obediently christened it the "Corsair Fleet," and among its enthusiastic members was Ernest Hemingway, who took to the seas off Cuba in his yacht, armed with a machine gun, hunting rifle, and hand grenades and visions of taking a U-boat singlehandedly. The press more accurately dubbed it the "Hooligan Navy." It did nothing but generate hundreds of false sighting reports.[30]

Roosevelt's other idea was to outbuild the U-boats' ability to sink ships. The mass-produced Liberty Ships that began pouring out of shipyards would become an American legend; by June construction had already accelerated to an astonishing pace, with sixty-seven launched that month, the fastest-assembled of them completed in sixty days start to finish. The German naval intelligence staff quickly revised its forecasts; it now estimated that 10 million tons of shipping would be added to the Allied merchant fleet by British and American shipyards in 1943, which meant that instead of 700,000 tons a month, Dönitz's U-boats would need to sink 900,000 a month to keep pace.[31]

Replacing men and ships faster than they could be destroyed was one way to win, and in some ways it was what *would* win the Battle of the Atlantic for the Allies, but it was a brutal calculus, a naval war of attrition at its most elemental. The military analysts Eliot Cohen and John Gooch would observe, "The undeniable resource shortages of early 1942 helped conceal the underlying problems of American ASW; the swelling tide of Allied ship production thereafter further obscured them."[32] The underlying problems were that American fliers and escort vessel commanders were not applying the lessons that the British, with the notable assistance of Blackett's operational research sections and Winn's Submarine Tracking Room, had already established for how to effectively protect convoys and destroy U-boats. In May 1942 Dönitz noted in his war diary, "The American airmen see nothing, and the destroyers and patrol vessels proceed at too great a speed to locate the U-boats or are not persistent enough with depth charge attacks."[33]

Baker had noted it, too. From January to June only 2 percent of attacks by U.S. naval forces resulted in the confirmed sinking of a U-boat; that was about a quarter the British success rate.[34] It had taken until April 14 for the U.S. Navy to sink its first U-boat: *U-85*, sunk by gunfire from the destroyer *Roper* off Cape Hatteras. Baker was convinced the problems never would be solved until command of the antisubmarine battle was centralized.

At the very end of April, Baker reached an agreement with King's staff that CominCh would publish an ASW manual and issue regular antisubmarine warfare bulletins. But still King refused to drop his basic belief in decentralized control. "This in no way restricts Fleet and Sea Frontier Commanders in the issue of bulletins within their own forces," the agreement emphasized. Moreover, the present arrangement in which CominCh, Atlantic Fleet, and the sea frontiers each retained their own ASW units would continue; even worse, Baker was to turn over to the sea frontiers the job of analyzing the operations of the Atlantic Fleet units temporarily attached to them, as soon as they were able to take on that duty themselves. The sea frontiers, for their part, were not to issue any doctrinal instructions to the air units attached to them; if they had any suggestions that applied specifically to antisubmarine air operations, they could "recommend" them to the separate army air commanders who were in charge of each different type of patrol plane and who were responsible for the training doctrine of the pilots under them. The one glimmer of hope in all of it was that CominCh had agreed to transfer to Washington the entire ASWORG team along with Baker to CominCh headquarters as soon as space was available in the Navy Department building.[35]

On June 24 Captain Baker sent a curt memorandum directly to Admiral King that minced no words about the woeful shortcomings of the arrangements. It began with an arresting statement:

> The Battle of the Atlantic is being lost.
>
> This is due to:
>
> (a) Lack of training by our A/S forces.
> (b) Lack of unified control of effort, i.e., rigid organization requiring conferences prior to taking action.
> (c) Insufficient vessels and aircraft.

Baker in his suggested remedies zeroed in on the two cruxes of the matter: the lack of proper attack procedures and instruction and the still fragmented command structure. District commandants, he pointed out, were distracted by local administrative duties. The Eastern Sea Frontier, the Atlantic Fleet, and CominCh ASW units all duplicated and overlapped one another; no one authority was responsible for training, planning, intelligence, and doctrine. Intelligence on the whereabouts of the U-boats was not getting to the tactical units that needed it. It could take weeks or months of bureaucratic turf battles to transfer ships or aircraft to a new sector as the enemy activity shifted around the Atlantic. Local commanders each had their own ideas about antisubmarine tactics; no one was examining the accumulated experience across the entire theater to draw lessons about what really worked best, and then following through with consistent training for destroyer and aircraft crews.

Baker forcefully urged combining all of these separate ASW units and appointing a single commander of anti-submarine warfare in Washington, where he would have direct access to the Convoy and Routing Section and other central functions of the navy, including intelligence and communications. The new commander should be given the unequivocal authority to develop and enforce proper antisubmarine training and procedures throughout the Atlantic theater. Subordinate commanders expected to do the actual fighting should "have no other duties than *fighting*."[36]

Baker then presumably held his breath and waited for the explosion.

BAKER'S OTHER PROBLEM WAS his rapidly growing team of scientists, who refused to behave as expected. No one in the navy seemed to know how to handle them or what to do with them. As Morse observed, the regular navy men at first could not even figure out whether the scientists were the equivalent of officers or ordinary seamen: "They didn't know whether to ask us or order us to do things." The scientists, in any case, rebelled almost immediately, Morse recalled:

> Almost as soon as we arrived . . . we were shown a room full of reports of all actions by or against enemy submarines, real or imagined. I suppose we were expected to file quietly in, to studiously digest all the reports, and once in a while to emerge to deliver some oracular pronouncement, which would then be implemented by the officers—provided they agreed with us. Our reaction to this unspoken assumption was unanimous, although we hadn't had the prescience to discuss it beforehand. We looked at a few reports and talked to some of the officers who had participated in U-boat sightings and attacks. And we said we wanted to think about the problem before we started to read. It must have seemed like procrastination to the officers, but they weren't sure how far they could order us around.[37]

Morse's team then "went into a one-week huddle" while they worked out, from first principles, a mathematical theory of submarine search. They were aiming to come up with a set of equations that would clarify what the key factors were that determined how often a submarine would be found by an aircraft conducting a visual or radar search over the ocean, or by a surface vessel scanning the water below with sonar. With that mathematical model, they could then start making recommendations for how much air coverage was needed to attain a given probability of finding a U-boat, what the optimal search patterns were, and other basic tactical and operational questions. ASWORG Report No. 1, "Preliminary Report on the Submarine Search Problem," was completed on May 1, 1942.[38]

Only when they were done did they compare their model to the data in the attack reports. One problem was immediately clear: the reports, which were filed after every possible submarine contact by a ship or plane, did not contain much of the quantitative data that was really needed to address the basic questions of search effectiveness. But even the data that was available agreed only poorly with the theoretical predictions from their model of how often a U-boat should have been sighted. "Our reaction to this disheartening discovery was just as unanimous, and just as visceral, as our first reaction," Morse wrote. Audaciously, "We believed our theory. We didn't believe the reports."[39]

The scientists went to Baker and explained that they wanted to go to the antisubmarine bases themselves, talk to the pilots, go out on some patrols if possible, and see if they could find out what was going on while also starting to collect more accurate data of the kind they already knew they needed. Baker countered by suggesting that officers could be brought in from time to time to the office to be interviewed. Morse held firm. Finally Baker agreed to see if he could get permission for the scientists to make some field visits. It was part inspection trip and part missionary work, Morse realized, and he knew if they blew it the scientists would not be permitted in the field again. As he would write later in a procedure manual on organizing an operations research section, a scientist on a field visit had to be free to ask questions and poke his nose into every aspect of the actual operations, but at the same time had to behave with "considerable delicacy" to avoid creating the suspicion on the part of the base commander that he had been sent as "a sort of spy" to check up on things. The best way to avoid trouble, Morse recognized, was to get the base commander on their side from the start by promising him that any recommendations or reports ASWORG produced would go directly to him for approval and distribution.[40]

The other way was to pick his first batch of field representatives with considerable attention to their personalities. Morse set the rules down in a mimeographed set of instructions that he made required reading. "To be read by each member at least once a month" stated the cover sheet. Morse admonished his staff that an ASWORG member's "usefulness in his job will depend at least as much on his tact and trustworthiness as on his scientific ability":

Our job is to *help* win the war, not to run it ourselves. We are novices at a task which has been worked at and thought about for many years. Our sole

> value is due to our specialized scientific ability. . . . We begin to be useful
> when we can combine with our scientific training a practical background
> gained from contact with operating personnel. This practical background
> can only be obtained when the operating personnel trust us and like us.

Scientists did not as a rule often care if anyone liked them, but Morse hammered at the point. Military rules and traditions might seem arbitrary, illogical, even idiotic, but if the scientists flouted them—security procedures were an obvious point of possible conflict—they would be finished before they even started: "A single slip by one of us could destroy the usefulness of the whole group."[41]

In the first week of April, Morse's deputy William Shockley had already made a preliminary trip to bases in Norfolk and Langley Field, Virginia, where he spoke to several B-18 pilots and flew on one patrol over Cape Hatteras, which he noted was "simply littered with wrecks and oil from wrecks" of torpedoed tankers and freighters. By June, Philip J. McCarthy, a mathematician, was stationed at Eastern Sea Frontier headquarters, and Arthur F. Kip, a physicist, at the Gulf Sea Frontier HQ in Miami; the next month another mathematician, Robert F. Rinehart, was sent out to the Caribbean Sea Frontier in San Juan, Puerto Rico.

Their immediate aim was to get better data coming in on a routine basis. The scientists posted in the field made a point of explaining their work, what they were trying to do, and what kinds of reports they needed. As they suspected, no one had paid much attention to the matter previously. After Kip gave his spiel to a group of pilots, one said, "Hell, I didn't think anyone ever read those damned reports."[42]

Meanwhile ASWORG had moved its remaining staff from Boston to Washington in June. The Navy Department headquarters building, known familiarly as "Main Navy," was a First World War–era monstrosity, intended originally as temporary office space to cope with the nation's sudden military mobilization. It was plunked down right smack on the Mall, along with an identical building for the Army's Bureau of Ordnance known as the Munitions Building. They were sprawling comb-shaped structures whose teeth formed multiple parallel corridors; the back of each comb faced Constitution Avenue while the teeth projected nearly all the way to the Reflecting Pool, which stretched east from the Lincoln Memorial. Franklin Roosevelt, as assistant secretary of the navy in 1917 and in his own mind an architectural authority, originally proposed putting the Navy Department building on the Ellipse, right in front of the White House, on the theory that it would thus be such an eyesore that "it would just *have* to be taken down at the end of the war." It hadn't worked out that way. President Woodrow Wilson objected that the sawing and hammering would disturb his concentration and Roosevelt reluctantly agreed to move the site a half mile down the Mall, to Potomac Park. The temporary buildings became permanent just as Roosevelt feared; in 1941, he remarked at a press conference that "it was a crime for which I should be kept out of heaven, for having desecrated the whole plan of, I think, the loveliest city in the world." (The last of the "temporary" structures was finally torn down in 1970.)

The ASWORG staff was at first accommodated in one of the even more temporary "T Buildings" the navy occupied across the street, but by the end of the summer they were installed on the third deck of Main Navy, Rooms 4303 to 4313.[43] Even the main building was dreary, distinctly un-air-conditioned, with water-stained ceilings, creaking doors, and walls and floors that would shake when anyone walked down the corridor, setting the bottles in the Coke machines rattling. But in 1942 it was undeniably where the seat of power of the U.S. Navy was to be found.

Washington was bursting at the seams with an influx of war workers and finding a place to live was a legendary trial, the source of countless personal sagas and black-humor jokes. Morse was commuting to Washington from Boston each week and he and Shockley ensconced themselves in a suite at the posh Hay-Adams Hotel near the White House; George Kimball, a former postdoctoral fellow of Morse's at MIT who had joined ASWORG at the end of July as the senior man in the Washington office, in charge of physical research and editing the reports, joined them there. Later Shockley took a room at the University Club on 16th Street, which had been a rather stodgy and creaky establishment until the wartime boom suddenly reversed its sagging fortunes and a long waiting list grew for membership.[44] The junior scientists were left to scramble to find a room with all the other clerks, typists, and other new arrivals who had suddenly doubled the city's population, to over one million, from just a year before. Some rooms were available in private homes after the city suspended zoning rules that prohibited renting out single bedrooms in residential neighborhoods; those less lucky doubled or tripled or quadrupled up in one-room apartments. A city housing inspector reported finding "Dickensian" conditions in many rooming houses. NEWCOMERS DISCOVER PRIVATE BATHS WENT OUT WITH HITLER, read a headline in the *Washington Post*.[45]

One of the first studies the ASWORG scientists began with their new trove of data was the optimal search and convoy escort patterns for patrol aircraft to fly. The actual sighting data allowed them to relate the factors that affected the visibility of a surfaced U-boat from the air—the aircraft's height, meteorological visibility, distance from the target—to its probability of being seen. They then worked out the most efficient patterns for an aircraft to orbit a convoy to sweep the maximum buffer area around the escorted ships. Shortly after arriving in San Juan, Rinehart obtained permission to have the local squadrons test the escort plans, which he had had a large part in devising. After some more tinkering, they were issued as an official U.S. Fleet doctrine, remaining in force throughout the war.

Another early ASWORG study showed that the wide discrepancy—as much as a factor of seven—in the number of flying hours per U-boat sighted by different air units was largely the result of inefficient overflying of the same space by the poorer-performing squadrons. Early studies also employed the sighting theory to develop what the scientists termed "gambit" tactics: a U-boat that quickly escaped by submerging when first sighted could sometimes be lulled into resurfacing after its crew was convinced the patrol plane had abandoned the search. The scientists calculated how long the plane should stay away from the area after the first sighting and then where to focus the patrol to have the best chance of intercepting the boat when it resurfaced.[46]

Whether it was an independent discovery or simply cribbed from the earlier British finding, ASWORG achieved its most attention-getting early success by recommending a change in the depth setting of air-delivered depth bombs from 75 to 25 feet. In his memoirs Morse attributed the discovery to Shockley, though at some point the American scientists definitely had in their hands a copy of E. J. Williams's Coastal Command ORS Report No. 142, which had reached the same conclusion a year earlier. In any case, within two months of ASWORG's Report No. 11 on August 2, 1942, "Aircraft Attacks on Submarines," the shallower depth settings were producing the same dramatic results they had in Britain. That success "gave us the beginnings of a reputation," Morse said.[47]

The first ASWORG report had been typed up like a college term paper but their subsequent publications started appearing with blue covers and professional hand-block-lettered titles and filled with sharply rendered maps and colored diagrams, the contribution of the architect who had joined the group, Donald A. MacCornack.[48] A few of the reports also sported whimsical cartoons on the back cover depicting a bare-breasted mermaid turning her back on Adolf, perched atop a U-boat with arms outstretched. The reports still had a limited distribution and Morse was careful to make sure not to step on any toes at CominCh; copies were circulated to ASWORG members in the field but it was up to Admiral King's staff to decide if anyone else in the navy received one.

His discretion paid off. Despite King's aversion to dictums from on high instructing subordinate commanders on matters of "how," nearly all of ASWORG's study results from its first few months of existence were shortly thereafter incorporated into a comprehensive "Tentative Doctrine for Anti-Submarine Warfare by Aircraft" being drafted directly by the CominCh staff. The document specified a 25-foot setting for depth bombs, attacking submarines only that had been submerged for less than 30 seconds, a tight stick spacing of 40 to 60 feet, and laid out the detailed aerial escort plans and "gambit" procedures developed by ASWORG. The whole tenor of it was a radical departure. It was simple and to the point; but it reflected in virtually every detail the scientific conclusions of men who had been with the navy for a few months rather than the traditional views of officers going back decades.[49]

9

Closing the Gaps

SINCE THE START of the war, Patrick Blackett had been living in a flat in Tufton Court, just a few hundred feet from Westminster Abbey and the Houses of Parliament. In 1942 he moved to the calmer and leafier town of Ruislip, a medieval parish to the northwest of the city that had been developed into a commuter suburb around the turn of the century with the extension of the Metropolitan Line of the London Underground. Much of the land of the parish was owned by King's College, Cambridge, which decided to sell it off for development; a member of the Royal Society of Arts picked out which ancient buildings were worthy of preservation (the Great Barn, which dated from the thirteenth century, the local public house) and the rest of the town was leveled to make way for new suburban villas. Blackett's new home overlooked the ancient parish common, kept as open parkland.

At the Admiralty, Blackett almost immediately found himself plunged into a controversy that tested where impartial scientific advice ended and the more fraught currents of grand strategy and politics began. The shifting of Dönitz's U-boat offensive to the easy pickings on the east coast of the United States through the winter and spring of 1942 had set off a scramble to find additional antisubmarine air forces. But there was a larger strategic question that had been growing for some time—and which would intensify through 1942 as the U-boats returned to their familiar haunts in the mid-Atlantic—about the most effective allocation of air power. By midsummer some 300 planes had joined the antisubmarine fight along the American coast; together with the crucial and continuing improvements in training and doctrine and the institution—at last—of an interlocking convoy system along the entire length of the coast it spelled the end of the "Second Happy Time."

The American aircraft hastily diverted to antisubmarine squadrons in the first half of 1942 included some 140 U.S. Army bombers, but nearly all of these were light and medium twin-engine planes such as the obsolescent B-18 and A-29 Hudson.[1] It had not been a very painful sacrifice for the Army Air Forces. The larger fight now about to be rejoined was over the allocation of the advanced, long-range, four-engine planes like the B-17 Flying Fortress and the B-24 Liberator now starting to roll out of factories across the American heartland.

The air force had them designed and built with one purpose in mind, to take the war directly to the enemy in a crushing application of strategic air power. The bomber barons were absolutely convinced that the way to win the war was to keep building up a mighty strategic force of heavy bombers that could attack the German heartland directly. Major General Carl Spaatz, commander of the U.S. Eighth Air Force and later of all U.S. strategic air forces in Europe, declared on more than one occasion that strategic air power would make an invasion of the continent and the conquest of Germany by ground troops simply unnecessary: the war would be over by then.[2] The American airmen were not so persuaded by Douhet's theories about the collapse of civilian morale through the bombing of enemy population centers. But they had worked out a detailed scheme of targeting vital war industries, transportation hubs, and fuel supplies which in their view would even more surely bring about the disintegration of the enemy war machine, by precision bombing carried out with vast air armadas flying at 30,000 feet over the enemy's territory.

The impatience in Britain to find some way to lash back at the Germans for the pounding the Luftwaffe had visited upon British cities and towns gave an extra impetus to getting the Allied strategic air campaign under way. When Churchill visited devastated neighborhoods of London's East End during the Blitz, walking amid piles of rubble left from the German bombers, crowds greeted him everywhere with the same cry: "We can take it, but give it 'em back!"[3] Churchill frequently asserted that the fight against the U-boats was the only theater where Britain could now lose the war; he also had come to the conclusion that the strategic air campaign was, at least possibly, the only one where the Allies could win, certainly the only way they could get in some blow against the enemy while the massive buildup for a landing on the European continent slowly proceeded. As early as July 1940 Churchill observed in a minute to his minister of aircraft production:

> When I look round to see how we can win the war I see that there is only one sure path. We have no Continental army which can defeat the German military power.... But there is one thing that will bring him back and bring him down, and that is an absolutely devastating, exterminating attack by very heavy bombers from this country upon the Nazi homeland. We must be able to overwhelm them by this means, without which I do not see a way through.[4]

With Russia's entry into the war a year later, the strategic air offensive gained yet another urgent rationale. It was the only way to aid their embattled ally until a second front in Europe could be opened against the Germans. In February 1942, Churchill got the "vigorous" commander he had been seeking to take charge of the air campaign at RAF Bomber Command. "Victory, speedy and complete, awaits the side which first employs air power as it should be employed," Air Marshal Arthur Harris told Churchill that spring. Harris had no time for what he termed the "panacea mongerers" who advocated half measures and circumscribed attacks on selected targets. They, he scoffed, believed you could "send a bomber to pull the plug on Hitler's bath so he would die of pneumonia." Invoking the terrible casualties suffered in the trenches of the First World War, Harris with equally brutal sarcasm dismissed objections that bombing cities was immoral. Anything that would shorten the war, he insisted, would save lives, and a war waged decisively by strategic air attack was an infinite improvement over a war of stalemate slogged out by "morons volunteering to get hung up on the wire and shot in the stomach in the mud of Flanders."[5]

Harris wasted no time putting his ideas into practice. On May 30, 1942, the new Bomber Command chief assembled a 1,000-plane force, pulling several hundred bombers from training units to do so, and ordered them to strike the German city of Cologne. The planes dropped 1,400 tons of bombs, two thirds of the tonnage in the form of incendiaries, over the space of two and a half hours. The resulting fires burned 600 acres of the city to the ground.[6]

Churchill would later have doubts; a year later, viewing a film of Allied bombing raids, he leapt up from his chair and exclaimed, "Are we beasts?

Are we taking this too far?" After the Allied firebombing of Dresden in February 1945 Churchill tried to distance himself even further from the policy (leaving an unrepentant Harris "the willing fall guy," in the words of the historian John Buckley). Even in the spring of 1942, the prime minister had backed away a bit from his earlier view that air power alone could win the war, cautioning Harris that he "must not spoil a good case by overstating it" when Harris insisted on making that argument in one memorandum. But Churchill was nonetheless convinced that Britain and the United States had to find some way to come to grips with Hitler without delay, and strategic air power was the only immediate avenue open to them. The Allies had to show that they were prepared to match the Nazis with equal toughness—and where necessary equal brutality, Churchill believed.[7]

The return of the U-boats in midsummer 1942 to the convoy lanes of the Atlantic underscored, however, that there was a competing mission for long-range bombers which might be as important if not even more vital to the war effort. Allied aircraft flying from bases in Newfoundland and Iceland could now cover much of the Atlantic. Yet there remained a 500-mile-wide "air gap" in the middle of the ocean that lay beyond the 600-mile reach of even the longest-range shore-based bombers and flying boats in the anti-submarine squadrons. The crews of American merchant ships already were calling the stretch of deadly unprotected ocean "torpedo junction." The only way to close the gap was with the same aircraft that the British and now American air forces were counting on to launch their all-out strategic air campaign against Germany.

The renewed threat to the Atlantic convoys was exacerbated by a dire change in fortune that had meanwhile taken place on the code-breaking front. On February 1, 1942, U-boats in the Atlantic began using a new version of the Enigma machine that had four rotors in place of the previous three, adding another multiple to the mathematical security of the system. The bombes were now useless for breaking the U-boat traffic. Because the U-boats had operated singly in their onslaught along the American coast, the full impact of this loss of intelligence was not immediately felt for the first half of 1942; the U-boats were not forming wolf packs and thus were sending and receiving few radio signals that would have yielded usable intelligence even if they could be broken and read. The main value of reading the U-boat radio traffic was to be able to know where the U-boat patrol lines were being formed, so as to reroute convoys around them.

With the Battle of the Atlantic returning to its familiar patterns in the summer of 1942, the loss of this vital intelligence source was a grievous blow. Worse, though the Allies were woefully unaware of the fact, the B-dienst had almost completely solved a new and supposedly more secure code that the British, American, and Canadian navies had begun using the previous fall to coordinate convoy movements. By March 1942 the Germans were reading as much as 80 percent of the traffic sent in Naval Cypher No. 3, frequently giving them current reports on the locations and sailing plans of convoys.[8]

In both Britain and the United States the air staffs controlled the allocation of new bombers and the submarine hunters of RAF Coastal Command and U.S. Army Air Forces I Bomber Command were not high on their priority lists. On February 14, 1942, Admiral King had formally requested the transfer of 400 heavy B-24 Liberators and 900 medium B-25 Mitchell bombers to the navy from current production, but the AAF shot that down at once. "There are no heavy or medium bombers available for diversion to the Navy," came the reply. The British Air Staff similarly told Coastal Command that Bomber Command had first claim on all new production of Lancaster bombers, the British counterparts of the four-engine Liberator that were just beginning to arrive.[9] As of mid-May, Coastal Command's antisubmarine squadrons had only 44 long-range aircraft, and only 18 of those were the Liberators and Catalina flying boats that had an effective patrol radius of 600 miles from shore; the remaining 26 were Sunderlands with a radius of about 450 miles. The other 198 aircraft available were all old medium-range Hudsons and Whitleys, which at best could reach a little more than 300 miles.[10] There was a crucial need both for additional long-range planes and for new "very long range," or VLR, versions of the largest planes that could close the air gap altogether. VLR aircraft could be produced by stripping some of the armor plate and gun turrets out of Liberators and using the weight saved for extra fuel tanks.

Late that month Admiral John Tovey, commander-in-chief of the Home Fleet, sent a scathing memorandum protesting the "absolute priority" accorded Bomber Command in the allocation of new aircraft. "Whatever the results of bombing of cities might be," he wrote, "it could not of itself win the war, whereas the failure of our sea communications would assuredly lose it." It was difficult to believe, the admiral later added, "that the population of Cologne would notice much difference between a raid of 1,000 bombers and one by 750." Tovey went so far as to urge the Board of Admiralty to resign en masse unless the navy were given the aircraft it needed to carry on the war against the U-boats.[11]

BEHIND THE TUG-OF-WAR between the navy and the air force for control of long-range aircraft, Blackett and Tizard were locked in a parallel war with their old nemesis Lindemann over the scientific evidence underpinning these competing strategic objectives. Since resigning from the Air Ministry's advisory committee after his showdown with Lindemann over the German beams in June 1940, Tizard had become a scientific adviser to the Ministry of Aircraft Production. Tizard had been skeptical for some time about the priorities being given to heavy bomber production. In August 1941 Lindemann had assigned a member of his staff, D. M. Butt, to analyze 650 aerial photographs taken by RAF bomber crews on their raids over Germany. The results were appalling and astonishing: only one in five crews who thought they had hit their assigned target had gotten within even five miles of it. "The war is not going to be won by night bombing," Tizard wrote to Air Marshal Sir Wilfrid Freeman, the chief executive of the Ministry of Aircraft Production, on December 24, 1941. Yet, he added, the current bomber production program "assumes it is."[12]

Early in the new year the Ministry of Home Security enlisted Solly Zuckerman and J. D. Bernal to undertake a quick scientific survey of the effects of the German bombing on the British cities of Hull and Birmingham. On February 17, 1942, Bernal came by to give Tizard a summary of their initial findings, which only further confirmed Tizard's doubts that bombing would ever do much of anything of real strategic value. Zuckerman and Bernal had assembled two teams of about forty people each to conduct a thorough block-by-block "bomb census," tabulating physical damage to buildings, injuries and fatalities, effects on factory production, and overall morale of the civilian population: they covered everything, said Bernal, down to "the number of pints drunk and aspirins bought." In neither town, they found, "was there any evidence of panic." Nor had worker productivity suffered. The only reduction in industrial production that had occurred was the direct result of physical damage to plants. Moreover, the actual number of casualties inflicted was remarkably small, given the 717 tons of bombs the Germans had dropped on the two towns during the period examined.[13]

Blackett talked to Bernal around the same time and did a few pages of penciled arithmetic. In the ten months from August 1940 to June 1941, the Luftwaffe had dropped 50,000 tons of bombs, an average of 5,000 a month. The number of civilian deaths was 40,000, or 4,000 a month; that worked out to 0.8 persons killed per ton of bombs. Experiments Zuckerman had ear-lier carried out using monkeys and goats to test the blast and concus-sion effects of explosions found that British bombs were about half as effective as German bombs of the same weight, which had a thinner casing and more explosive. Also, because British bombers had a greater distance to fly to reach their targets, the resulting navigational difficulties meant that British aircrews were—as Butt's photographic analysis had shown—at best half as likely as the Germans to drop their payloads on built-up areas of their target cities. In the eight months prior to February 15, 1942, RAF Bomber Command had dropped 2,000 tons of bombs a month; that worked out to $2,000 \times 1/2 \times 1/2 \times 0.8 = 400$ German civilians killed a month. Blackett noted that this was almost exactly the same as the number of British airmen killed in the 728 planes shot down carrying out the raids over that same period. (After the war, when the actual figures on German casualties became available, Blackett found that even his estimate had been optimistic; only 200 civilians had been killed a month by the British bombing raids in 1941.) He also calculated that the total decline in industrial production from the air attacks on Britain was less than 1 percent: factory output for the country in April 1941 had been affected more by the Easter holiday than by the German bombs that fell during that month.[14]

Blackett showed his figures to Tizard the same day that Tizard met with Bernal, February 17. The next day Tizard sent a note to the minister of air-craft production:

> I say emphatically as a conclusion, that a calm dispassionate review of the facts will reveal that the present policy of bombing Germany is wrong; that we must put our maximum effort first in destroying the enemy's sea com-munications and preserving our own; that we can only do so by operating aircraft over sea on a very much larger scale than we have done hitherto, and that we shall be forced to use much long range aircraft. . . . The heavy scale [of strategic bombing] will only be justified and economic at the concluding stages of the war when (or if) we are fortunate enough to have defeated the enemy at sea and to have command of it. Until that time is ripe, everything is to be lost by concentrating on this bombing offensive instead of by concentrating on the sea problem.[15]

A few days later Tizard sent a letter to Freeman. He reiterated the key point that the Air Staff had not even begun to grasp the scale of bombing needed to yield even minimal results. The air marshal's reply made it clear that Blackett and Tizard had thrust themselves into a political arena where facts that conflicted with policy would simply be ignored, and those who raised them would be regarded as disloyal:

> My dear HT, I have read your personal letter to me about bombing, and my first reaction is that you have been seeing too much of Professor Blackett. . . . Is not Blackett biting the hand that fed him?

Not long after that Tizard arrived at his desk one morning to find a transcript of a German radio broadcast that had denounced the latest British bombing attacks. Attached was a note from Freeman: "Like you, the Germans are very anxious for us to stop raiding their towns."[16]

Lindemann sided strongly with the Air Staff. Faced with Zuckerman and Bernal's conclusive finding that bombing did not inflict very many casualties, Lindemann decided with magnificently circular reasoning that bombing must therefore be effective for *other* reasons. He proposed that the real purpose of strategic bombing was to "dehouse" the enemy workforce; this would cause production and morale to collapse. On March 30, Lindemann—now Lord Cherwell, having been rewarded by Churchill with a peerage—circulated a memorandum to the War Cabinet in which he concluded that one third of the entire population of Germany could be "turned out of their house and home" by an Allied strategic bombing offensive that employed a force of 10,000 heavy bombers to target the fifty-eight German cities with a population of 100,000 or more:

> Investigation seems to show that having one's house demolished is most damaging to morale. People seem to mind it more than having their friends or even relatives killed. At Hull signs of strain were evident, though only one-tenth of the houses were demolished. On the above figures we should be able to do ten times as much harm to each of the 58 principal German towns. There seems little doubt this would break the spirit of the people.[17]

It was only after the war that Zuckerman and Bernal discovered how completely Cherwell had misrepresented their findings about civilian morale in Hull and Birmingham. Their study, which Cherwell had asked for a preliminary report on, had concluded almost exactly the opposite of what he was arguing in his memorandum.

Blackett went to see him. Cherwell, he recalled, did not try to dispute Blackett's calculations about the direct effects of bombing. But it was clear he was once again enraptured with one of his pet ideas. It was the first time Blackett was fully confronted with this side of the Prof's character, "his almost fanatical belief in some particular operation or gadget to the almost total exclusion of wider considerations," Blackett later wrote. "Bombing to him then seemed the one and only useful operation of the war. . . . Never have I encountered such a fanatical belief in the efficacy of bombing." Cherwell went so far as to insist that *any* diversion of aircraft to *any* other purpose would be a "disastrous mistake."[18]

When Cherwell's paper proposing the air campaign to "dehouse" the German population was circulated to the Admiralty on April 8 for official comment, Blackett was asked by Admiral Godfrey, the director of naval intelligence, to prepare a "full factual examination of this problem" in his capacity as chief advisor for operational research. Blackett in his response reiterated the calculations he had shown Tizard earlier and also presented some new data concluding that if even a quarter of the 4,000 bomber sorties which could be flown each month by the summer of 1942 were diverted instead to antisubmarine operations, it would save a million tons of Allied shipping a year.[19]

Tizard and Blackett also realized that Cherwell had hugely overstated—by 600 percent—the quantity of bombs that could be delivered against German cities even if the entire forecasted production of heavy bombers over the following eighteen months was devoted exclusively to the job. Cherwell had in effect assumed that every bomber scheduled to be produced would be instantly available and by the end of the eighteen months would already have carried out the twenty bombing missions that were the average operational life of each aircraft. It was a ridiculous arithmetical mistake. Again, subsequent history showed that Blackett and Tizard had been too cautious in their critique: Cherwell had overestimated not by 600 percent but by more than 1,000 percent.[20]

It all fell on deaf ears. "I do not think that, in secret politics, I have ever seen a minority view so unpopular," C. P. Snow would write of the episode.[21] Blackett made a wan joke that if anyone at the Air Ministry during this time "added two and two together to make four," suspicions would arise that "he has been talking to Tizard and Blackett" and was "not to be trusted."[22]

Blackett began keeping a file of sardonic observations about official attitudes and the military mind; a quotation he had run across in a Spanish anarchist newspaper he thought might perfectly summarize the reaction he had received to one study: "Let us have no more of these miserable statistics, which only paralyse the brain and freeze the blood."[23] He also filed away an anonymous and extremely incendiary paper he had been sent around this time that was full of an insider's scathing appraisal of Churchill; it noted how Churchill used his dazzling oratorical skills to avoid hard facts, how he browbeat his military advisers to agree with his decisions and then used them as scapegoats when things went wrong, how he worried subordinates about small details and demanded immediate answers or reports which frequently turned out to be wrong because of the haste with which they were prepared, how even his wide knowledge and experience was often more a danger than an asset: "His judgment is very unreliable and in spite of his study and past service he has never understood naval warfare, in which his interest is extremely doubtful."[24]

The hard feelings—and a sense of failure—about the fight over strategic bombing haunted Blackett the rest of his life. As he wrote in an article in *Scientific American* in 1961, it was his one great regret about his wartime service:

> So far as I know, it was the first time a modern nation had deliberately planned a major military campaign against the enemy's civilian population rather than against his armed forces. During my youth in the Navy in World War I such an operation would have been inconceivable. . . . If the Allied air effort had been used more intelligently, if more aircraft had been supplied for the Battle of the Atlantic and to support the land fighting in Africa and later in France, if the bombing of Germany had been carried out with the attrition of enemy defences in mind rather than the razing of cities to the ground, I believe the war could have been won half a year or even a year earlier. The only major campaign in modern history in which the traditional military doctrine of waging war against the enemy's armed forces was abandoned for a planned attack on its civilian life was a disastrous flop, and I am sure that Tizard felt the same way. If we had only been more persuasive and had forced people to believe our simple arithmetic, if we had fought officialdom more cleverly and lobbied ministers more vigorously, might we not have changed this decision?[25]

With the War Cabinet's approval for a campaign "focussed on the morale of the enemy population," Harris proceeded to put into practice his conviction that the only way to destroy *anything* in a city was to destroy *everything* in it, by setting huge conflagrations with tons of incendiary bombs. Harris was disappointed that two subsequent 1,000-plane raids he mounted following the attack on Cologne—the Ruhr cities of Bremen and Essen were hit in June 1942—were not as successful. Even the Cologne raid, he felt, ought to have done more destruction. But then the average bomb load of his planes was only a ton and a half; most of the force available to Bomber Command was still two-engine medium bombers. Harris was confident that once the new heavy bombers began to arrive in numbers, his program for "the elimination of German industrial cities," as he put it, would swiftly be accomplished.[26]

WITH BLACKETT'S DEPARTURE to the Admiralty in December 1941, E. J. Williams had taken over as head of the Coastal Command Operational Research Section, and within a few months he believed he had figured out a way to dramatically increase the effectiveness of the antisubmarine air forces even without prising a single new airplane away from Bomber Command. Based on some preliminary theoretical calculations, Williams sent Air Marshal Joubert a memorandum proposing an urgent study of maintenance and flying schedules of the aircraft in the command, and in June 1942 assigned a single member of his now twelve-person ORS staff, Cecil Gordon, to the task.[27]

Gordon, leaving behind his drosophila in Aberdeen, had just joined the unit that summer. He was pudgy, Jewish, abrasive, and communist. "Definitely not officer material," noted his Admiralty file. He had a high-pitched, grating voice, an awkward walk, a nervous fidget, and an unkempt appearance that was a source of wonder even to his left-wing scientist friends, not exactly known for fastidious dressing themselves. Once Gordon tried to donate some of his old clothes to a secondhand shop: the shop rejected them.

Gordon had literally been a card-carrying communist, joining the party in the 1930s and resigning apparently only just before he was hired by Coastal Command. His Marxist convictions nonetheless remained deep and indelible. Growing up in an impoverished Russian immigrant family in South Africa—his father was a fabulously unsuccessful peddler of geegaws in the native areas—Gordon had been powerfully influenced at the University of Cape Town by the antiestablishment views of Lancelot Hogben, the head of the zoology department where Gordon studied and later taught as a lecturer. Hogben was a socialist, militant atheist, ardent feminist, and outspoken critic of eugenics and the theories of scientific racism. In 1931 Gordon followed Hogben to the London School of Economics, where his mentor had just been appointed professor of social biology; he then completed his Ph.D. in genetics at University College London under another famous left-wing biologist, J. B. S. Haldane, before rejoining Hogben at Aberdeen. Gordon was thoroughly steeped in left-wing politics during his time in London and Aberdeen. He met his future wife at a socialist student dance; he chaired a British-Soviet Unity Campaign; a story made the rounds that he had been seen outside the Finchley Road tube station, dressed in dinner jacket, peddling copies of the *Daily Worker* one evening while on his way to dine with his wife's wealthy parents.

He was also impatient, vociferous, opinionated, and domineering in conversations. His mind was always "temporarily one-track," a colleague remembered, whatever obsession of the moment taking over his thoughts and conversation "to the exclusion of everything else for the time being." He could be brilliant and fascinating, relating some novel idea he had pulled from his enormous range of reading. When the topic of his momentary obsession was a problem of work, however, he could be "monotonous," and when it was one of his many grievances, "a bore." Even his enemies and detractors, though, acknowledged that he had a fierce intellectual and moral honesty, which he applied as ruthlessly to himself as to others.[28]

Within three weeks of being assigned to the maintenance problem Gordon had worked out the basic mathematical concepts and had thrown himself into an experimental field test of the ideas he had devised. The calculations did not require more than algebra but were lengthy and incorporated vast amounts of data that Coastal Command had collected about the work of the maintenance shops: the number of man-hours spent on various tasks, the intervals between inspections, the average time of failure of various components.

Gordon's basic conclusions, though, were simple and stunning. First was that there were strangling bottlenecks in the maintenance chain: the availability of skilled labor was the limiting factor at several choke points, which meant that while some repair departments worked constantly others were left idled with nothing to do. The analysis suggested that maintenance crews were accomplishing only three quarters of the work they could have in the total man-hours they had available if the work were more rationally organized. It was the four-washtub problem writ large.[29]

But a far greater obstacle to increasing the number of flying hours per month, Gordon realized, had to do with the RAF's "serviceability" policy that set a goal of having 75 percent of the aircraft of every squadron ready for operational duty at all times. The counterintuitive mathematical result Gordon discovered was that by accepting a lower serviceability rate, the total number of monthly flying hours would *increase*. The best policy would in fact be to ignore serviceability altogether; it was the wrong quantity to be measuring or even paying attention to. A serviceability standard made sense for fighter defense squadrons, which had to be ready at any time to put as large a force in the air as possible with minimal warning. But for the squadrons of Coastal Command, which had a steady day-to-day mission of patrolling against a constant enemy presence, the insistence on keeping a large portion of aircraft ready at any time could only be met by holding back a significant number that could otherwise be flying useful missions.

The reductio ad absurdum of this policy, Blackett later observed, would be never to fly at all: that would achieve a serviceability rate of 100 percent. (Blackett drew an analogy to an incident he recalled a friend relating in the years between the wars: arriving at a rural train station one night, he approached the lone taxi waiting out front and asked to be driven to his destination. The taxi driver refused—explaining that a local police ordinance required one taxi to be waiting at the station at all times.) Gordon's numbers suggested that if all serviceable aircraft were flown every day when weather permitted, the serviceability rate would sink to 30 percent, but monthly flying hours would substantially increase. The other way to think about it was that to get the most out of a squadron's aircraft, the best plan was to fly enough to ensure that the maintenance shops were fully employed at all times. To get more flying hours, in other words, you had to *increase* the breakdown rate. That would mean more aircraft needed repair at any given time, but the total throughput of the maintenance shops would increase.

Implementing Gordon's scheme for what would come to be called "Planned Maintenance, Planned Flying" was a prime case where supreme tact was required; in its bare outline it sounded like nothing so much as a criticism of the efficiency of a unit and a slap at workers for sitting around and doing nothing. Gordon was hardly tactful. But his sheer force of intellect seemed to make up for it. It also helped that Churchill was right on top of his work. On July 5 the prime minister asked the Air Staff for a copy of Gordon's initial report and quickly requested the first lord of the Admiralty and the secretary of state for air to sign off on the proposal to put it to a test: Churchill especially warned that until Coastal Command fully implemented Gordon's ideas to increase the number of sorties per squadron, "there can be no case for transferring additional squadrons from Bomber Coastal Command." Churchill also, significantly, grasped at once the essential point that "it is true the standard of serviceability in Coastal Command will fall off if the aircraft make more frequent sorties," but that this was an acceptable trade-off.[30]

Coastal Command agreed to place one unit, No. 502 Squadron of Whitleys, under Gordon's mathematical direction as a test. For five months Gordon tracked the status of every aircraft in the squadron as the changes were implemented. The crux of the matter was the percentage of time each plane spent in each of four different states: flying; serviceable but not flying; being serviced; or awaiting service. As expected, the percentage of time a plane spent waiting to be serviced increased. But the increase in flying hours more than made up for it: flying hours nearly doubled.

In the process, the scientists poked into everything from inspection schedules to the time spent waiting for spare parts to arrive to the rate at which various components of an aircraft fail. One of their interesting discoveries was that for a large class of components, routine inspections were literally a waste of time: a part like a fuse or a spark plug lead was "just as likely to go wrong at one moment as the next" and inspections revealed little or nothing. In fact, when they started plotting the failure rates of various parts, they found that inspections in many cases *increased* breakdowns, apparently the result of disturbing components that had been working fine. The results produced by the rationalization of maintenance procedures and flying schedules in 502 Squadron were another of those operational research contributions that spoke for themselves. Joubert ordered the policy instituted throughout the command—and asked the ORS scientists to take charge of implementing it.[31]

The mathematician T. E. Easterfield, who joined the group later that year and was promptly grabbed by Gordon to help with the work, found that the general attitude among the aircrews was that the scientists "might be mad, but they got results." Easterfield thought the scientists were viewed as "licensed jesters" whose odd behavior was tolerated because they served a purpose. Gordon's one-track mind and tenaciousness were madder than most; but so, too, his results were commensurately greater than most. "Marvelously efficient chap!" remarked one RAF officer to Easterfield about Gordon—before adding, "But he was no gentleman!" To the RAF aircrews,

Gordon was known as Joad, a nickname they bestowed on him after C. E. M. Joad, the eccentric and discussion-monopolizing resident philosopher on the popular BBC radio talk show *The Brains Trust*.[32]

WILLIAMS AND BLACKETT WERE meanwhile working on another approach to tighten the noose around the U-boats. Since 1941 air patrols had been regularly flying over the Bay of Biscay, the roughly 300-mile-by-300-mile bottleneck that all of the U-boats had to transit between the French ports and their stations on the mid-Atlantic convoy lanes. Williams calculated that in the second half of 1941 every U-boat that put to sea on a patrol had a 30 to 35 percent chance of being attacked by Allied aircraft during either its outward or inward passage of the bay.[33] In response, the Germans had taken to transiting the bay submerged during daylight hours, surfacing at night to recharge their batteries.

Most of the Coastal Command aircraft by this point were equipped with 1.5-meter-wavelength ASV Mark II radar sets, but the results had been disappointing. The aerials for the radars were an ungainly sight, mounted on a row of four pylons perched along the spine of the planes' fuselages. The radar's detection range was not much better than what the naked eye could achieve by day. At night, the problem turned out to be not how far the radars could see ahead, but rather their *minimum* range. The powerful radio pulse sent out by the transmitter would instantly fry the circuitry of the sensitive receiver located right next to it, so the receiver was automatically switched off for a fraction of a second as each pulse went out. But this meant the system was unable to detect echoes bouncing off very close objects, which would arrive back at the receiver during the hundred thousandth of a second or so while it was still switched off. In practice, anything closer than three quarters of a mile was invisible. If the sea was choppy, waves on the surface would add a clutter of echoes that extended out even farther. The ASV planes could thus pick up on their radar screens surfaced U-boats at night at a range of as much as ten to twenty miles, but as they approached they would lose the radar contact at a point where the boat was still well beyond the range the eye could see at night; there was no way to follow through with the actual attack.[34]

One way to bridge the gap between the minimum radar range and the point of attack was to equip the patrol planes with powerful searchlights that could be switched on for the final attack run. The first of these needed their own Ford V-8 engine and 35-kilowatt generator, or a bank of batteries completely filling the bomb bay, to supply the required electricity. An inventive RAF officer assigned to Coastal Command headquarters, Squadron Leader Humphrey de Verd Leigh, came up with the idea for a more efficient carbon arc lamp that would need only seven 12-volt batteries, which could be trickle-charged from a small generator driven by one of the aircraft's engines. The whole package weighed less than 600 pounds.[35]

In a February 1942 paper, Williams had calculated that a balanced day and night force with 50 Leigh Light–equipped aircraft and an additional 100 long-range Catalinas or Liberators could increase the chances of intercepting each U-boat on its inward or outward passage to well over 50 percent.[36] He was also able to provide a reassuring calculation proving that the Germans had not yet taken the step of countering the Allied radar by equipping the U-boats with warning receivers that could pick up a radar signal, which would give them time to dive before the would-be attacker came within striking distance. The poor success of No. 502 Squadron that same month when it tried to carry out night radar attacks using dropped flares to illuminate the target area made some Coastal Command officers worry that the Germans already had deployed such a threat-warning receiver. But Williams showed that the number of sightings by day per flying hour for ASV and non-ASV equipped aircraft were still roughly the same.[37]

By June 1942 the first Leigh Light–equipped planes were operating and the number of nighttime attacks shot upward.[38] Again, the operational researchers were able to produce convincing calculations showing that the sighting rate was very close to the theoretical values expected, assuming the Germans still had not deployed a radar warning device.

The impact on German morale was instantaneous. The security of the night had been shattered: Dönitz ordered his boats to reverse procedure and return to surfacing by day, the idea being it would be a more even fight in daylight when the U-boat crews could visually spot an approaching airplane and possibly have time to submerge. By July the average number of U-boats at sea reached seventy and the return to the convoy attacks was keeping the tonnage total at a steady half million or so a month.[39] But it was a much tougher and grimmer fight than ever. On July 27, Dönitz made a radio broadcast to the German people unlike anything they had ever heard from a high official of the Third Reich. He told his listeners that the "harsh realities of submarine warfare" meant Germany must expect sacrifice and losses. Dönitz later explained that he was worried by the exaggerated hopes of easy victory that had been raised by a torrent of official propaganda and felt that a dose of caution was needed to prepare the public that, as he said in his broadcast, "even more difficult times lie ahead of us."[40] It was, thought the British Admiralty, a clear indication that Dönitz was planning to pour even more of his force into the convoy battle.

In September he issued new orders emphasizing that the war was now a fight for survival. The immediate impetus for the orders was a chaotic incident in which the British troopship *Laconia*, carrying 1,800 Italian prisoners of war from North Africa along with some British soldiers and women and children, was torpedoed 500 miles north of Ascension Island by *U-156*. The German submarine assisted the survivors, taking nearly 200 on board and towing four lifeboats while radioing a message that it would not attack any Allied vessel that came to assist. Dönitz approved, and requested Vichy French ships from West Africa to help in the rescue of the British passengers and Italian prisoners. Two days later an American B-24 from Ascension Island spotted the surfaced U-boat, and attacked. The German captain immediately ordered his passengers onto the lifeboats, cut them loose, and escaped serious damage by diving, but one of the lifeboats filled with Italian POWs was sunk. A few hours later the Vichy French ships arrived and picked up the remaining survivors.

It was a confused incident and it was unclear how much of Dönitz's humanitarian concerns were prompted by the consideration that one of his captains had torpedoed a ship full of Axis prisoners. But it became the pretext for his order of September 17 that was the final culmination of the progressive brutalization of the war at sea:

To all Commanders:

1. All attempts to rescue members of ships sunk, including attempts to pick up persons swimming and placing them in lifeboats, righting capsized lifeboats, or supplying provisions and water, must cease. Rescuing survivors contradicts the most primitive demand of war, to annihilate enemy ships and crews.

2. The order [previously issued] for bringing back commanding officers and chief engineers remains in force.

3. Survivors are to be picked up only in cases where their interrogation would be of value to the U-boat.

4. Be severe. Remember the enemy in his bombing attacks on German cities has no regard for women and children.[41]

At some point that fall Dönitz issued another order, which was formally reiterated the following year in a war diary entry that referred to this earlier directive, calling the attention of his U-boat commanders to the usefulness of torpedoing the rescue ships attached to convoys. "In view of the desired annihilation of ships' crews," Dönitz stated, "their sinking is of great value."[42]

The increased Allied air activity over the bay had given Dönitz a moment of uncharacteristic doubt about whether the war could be won at all. He noted in his war diary that "the numerical strengthening of enemy flights, the appearance of a wide variety of aircraft types equipped with an excellent location device against U-boats have made U-boat operations in the eastern Atlantic more difficult." He went so far as to admit that if the situation in the bay continued, it might lead to "a decline in the prospects of success of the U-boat war as a whole."[43] The pressure eased in late September with the arrival of the countermeasure the British scientists had long anticipated, a radar search receiver that could be used by the U-boats while transiting the bay. The German electronics industry had largely been commandeered by Göring's Luftwaffe to produce radios and other equipment for aircraft and had no spare capacity to develop or manufacture new devices for the navy, but Dönitz at last was able to find two French companies to do the work. The Metox receiver (named after one of the French firms) could detect the 1.5-meter signals from the British ASV Mark II radars at a range of over thirty miles. That was farther than the radar system could detect a surfaced U-boat. The first sets were rather crude, consisting of a wooden-framed cross-shaped aerial that had to be lashed to the conning tower and removed before the boat could dive. The crews dubbed it the Biskaykreuz—the Cross of Biscay. Once again the U-boats could make the perilous passage on the surface by night with little fear of being caught without warning. In October, Coastal Command's ASV squadrons detected only a single U-boat at night for the entire month.[44]

Despite that temporary setback, the experiment in bringing the fight to the enemy in the bay had yielded a wealth of data on which to lay future plans for a more decisive operation. In Coastal Command ORS Report No. 204, "Air Offensive Against U-boats in Transit," issued on October 11, 1942, Williams was able to calculate precisely how much flying time was required to sink a U-boat in the bay. He also was able to forecast, with what would prove to be remarkable prescience, the sequence of tactical and technological countermeasures that were likely to ensue as the war progressed.

And because by this point the chances of sighting a U-boat under any set of given circumstances could be calculated "with considerable accuracy," as Blackett would note, it would be possible to get almost instantaneous warning if the Germans shifted tactics or introduced a new technological countermeasure, just by monitoring the sighting rates attained by sub-hunting aircraft.[45] Regardless of the inevitable evolution of gadgets and tricks, the methods that Blackett and Williams had largely been responsible for developing and perfecting would keep the Allies abreast of them. In any case, the current advantage that the Metox receiver had given the U-boats in the bay would only last until Allied aircraft equipped with the new centimeter-wavelength radar arrived. The first of those were expected within months.

THE TEMPO AND INTENSITY of the thrusts and counterthrusts in the Battle of the Atlantic was matched by developments in the cryptographic shadow war throughout that fraught summer and fall of 1942. The B-dienst's success against the Allied Naval Cypher No. 3 had played a major part in Dönitz's decision to rejoin the convoy battle in the summer of 1942.[46] Key to making the wolf pack tactics work was knowing where the convoys would be ahead of time. Meanwhile, Bletchley's blackout in reading the Atlantic U-boat Enigma signals continued.

GC&CS's breakthrough the previous summer in the naval Enigma had been the British code breakers' one remaining ace in the hole in negotiations with their American counterparts. While offering to share the intelligence results they extracted from the German naval signals, Bletchley insisted on keeping control of the actual code-breaking process. The British had also evaded repeated requests from the U.S. Navy for blueprints of the bombes. Partly, it was a matter of ongoing British doubts about whether the Americans could be trusted to keep a secret. More important, however, was the Admiralty's keen awareness that whoever controlled the information could control how it was used operationally. Knowledge really was power in this case. But Britain's leverage was rapidly slipping: you could hardly call the game if you were out of chips. Everyone who was working on the problem recognized that the ultimate solution was to build a four-rotor version of the bombes: these could apply the same proven method that worked before (and which was still working to break daily traffic of the three-rotor Enigmas used on German air force and army networks) to crack the four-rotor Enigmas now employed on the Atlantic U-boat network.

The trouble was that British electrical and mechanical engineering was turning out to be unequal to British mathematical genius. To work through every possible combination of four rotors in a reasonable amount of time required the fastest wheel to spin at a dizzying 2,000 rpm. The engineers of the British Tabulating Machine Company had barely managed to get the electrical contacts on their three-rotor machines to work reliably at speeds of under 100 rpm. Work had been dragging for months on the new four-rotor bombes, and was getting nowhere.[47]

By the summer of 1942, the U.S. Navy's cryptanalytical unit (known only by its bureaucratic designation in the navy command structure, Op-20-G) concluded that what was clearly needed was a dose of good old American know-how, and shocked their British colleagues with the pronouncement that they were going to enlist American industry to build a few hundred of their own four-rotor machines to get the job done. British secret service officials fought a feeble rearguard action for a few weeks, then threw in their hand. Faced with the inevitable, Edward Travis, who had succeeded Denniston as Bletchley's director following the prime minister's ACTION THIS DAY directive, and Frank Birch, head of the naval section, departed for Washington in September to negotiate a formal agreement for "full collaboration" between GC&CS and the U.S. Navy on the naval Enigma problem. Later that fall the U.S. Navy signed a $2 million contract with National Cash Register to build the four-rotor bombes at its Dayton, Ohio, plant. In the end, NCR, assisted by 600 WAVES who were dispatched to Dayton, would build 125 of the machines at twice that total cost.[48]

None of this solved the immediate problem. The new machines could not possibly be ready before spring 1943, and the pressures for action were mounting daily. November 1942 saw the greatest monthly sinking of Allied merchant shipping by U-boats of the entire war, more than 800,000 tons. Churchill convened a new Cabinet Anti-U-boat Warfare Committee and began his usual hectoring from on high. On November 22, the Admiralty's Operational Intelligence Centre added a circumspectly worded but urgent plea, asking the GC&CS naval section if "a little more attention" might be paid to the naval Enigma problem.[49] Two days later a pile of documents landed at Bletchley Park like a deus ex machina.

Since the beginning of the blackout of the U-boat traffic in February 1942, Alan Turing and the other Bletchley mathematicians assigned to the naval Enigma problem had been trying to devise a way to break back into the system even without a four-rotor bombe. Through another feat of pure mathematical analysis, they quickly determined that the four-rotor machines had been designed to be compatible with the existing three-wheel Enigmas so that the U-boats could continue to communicate with networks that had not been issued the more secure version. When the fourth rotor was set to the "A" position, the machine acted exactly like a three-rotor machine. What was more, certain kinds of short signals were routinely sent by the U-boats with the fourth rotor set to this "neutral" position. These were mainly weather reports, which were encoded by a series of letters of the alphabet according to a printed weather codebook before being enciphered on the Enigma and radioed.[50]

The other promising discovery the Bletchley naval staff made was that the U-boat weather reports were regularly retransmitted for the benefit of the entire German navy by a powerful transmitter at Norddeich, on the North Sea coast, using a simple code that Bletchley had already broken. It was generally an easy matter to match up a Norddeich weather report with the U-boat Enigma message that had provided it; one clue was that the weather short-signal code allowed latitude and longitude and wind speeds to be specified only in round figures, so any Norddeich report containing wind speeds divisible by four or latitudes and longitudes given to a whole degree had almost surely come from a U-boat.[51] All that Bletchley needed now was the weather codebooks themselves to be able to construct a three-rotor bombe crib that it could solve with its existing methods. Having already mathematically reconstructed the internal wiring of the fourth rotor by analyzing the signal traffic, they would then be able to read all the U-boat traffic directly.

On October 29, 1942, a British patrol in the Mediterranean spotted a surfaced U-boat halfway between Port Said and Haifa, and four destroyers were ordered to head for the spot. Shortly after noon they made sonar contact, and for the next ten hours the warships waged a relentless depth charge attack against their hiding prey. The last salvo ripped a hole in the bow of the submarine, sending it hurtling to the surface and ending the fight. Survivors told British interrogators that they had counted 288 explosions as they lay beneath the surface.

Within minutes of breaking the surface, *U-559* was being boarded by three sailors from HMS *Petard* who had leapt into the water and swum across. With water already rising belowdecks, the three began grabbing armfuls of documents from the control room and captain's berth and carrying them up the ladder to pass to another boarding party that had meanwhile arrived alongside in a whaleboat. One of the officers had just ordered the men to come up at once when the U-boat suddenly plunged beneath the surface. Tommy Brown, a sixteen-year-old canteen assistant who had lied about his age to join the navy, was just able to jump free in time and avoid being pulled under by the eddying undertow of the diving U-boat. Able Seaman Colin Grazier and Lieutenant Anthony Fasson were carried down to their deaths. Brown was presented a medal—and a discharge from the navy for being underage.[52]

It was grudging recognition considering his act helped win the entire Battle of the Atlantic. Among the documents Brown and the others managed to rescue was the latest weather codebook and a list of Enigma settings used to transmit the short signals. After running the bombes for two weeks straight, Bletchley sent a teleprinted message to the Admiralty early on the afternoon of December 14 giving the positions of a dozen U-boats in the Atlantic. At 9:00 p.m. the day after Christmas the code breakers read the first current day's U-boat messages in almost a year. Some 39,000 U-boat decrypts would follow to the end of the war. The December 26 break came just hours too late, however, to save convoy ON 154, which had been located by a U-boat that same afternoon. Two days later a wolf pack sank nineteen freighters and two escorts in the convoy.[53]

But over the next two weeks Dönitz's U-boats in the Atlantic were unable to locate a single one of eight "expected convoys" that BdU had alerted them to based on the B-dienst's decodes of Allied convoy signals. Every one was successfully diverted around the waiting wolf packs. Moreover, the signals exchanged between the U-boats and BdU that were required to organize a wolf-pack attack, and which were now being read by Bletchley Park, frequently gave Coastal Command time to dispatch one of its small number of VLR aircraft on the 1,000-mile journey to protect a convoy before the attack could be mounted. There was, Blackett noted, "almost never less than twelve hours and seldom less than a day" of advance warning that an attack was imminent. The Enigma intelligence was a literal force multiplier, allowing the still severely limited VLR force to become forty times more effective than it otherwise would have been: Blackett calculated that a VLR aircraft sent out to intercept a convoy known to be shadowed by a U-boat successfully sighted a U-boat 1 time in 2.4 sorties, versus 1 in 100 when simply assigned to escort a convoy as a general defensive measure.[54]

Once again puzzling over the curious elusiveness of the convoys, Dönitz ordered a full-scale security investigation. There were now only two possibilities, he noted in his war diary: "(a) that the enemy has succeeded in temporarily breaking into our cipher data or (b) that in some place or other in our own ranks there is treason." The answer was staring him in the face. Investigators from the Naval Communications Service in Berlin combed through the signals the British and Americans were radioing their convoys and found several that should have rung alarm bells. On repeated occasions convoys had been sent signals ordering them to change course to evade U-boat patrol lines or correctly reporting the locations of U-boats with an accuracy that surpassed what should have been possible by radio direction finding alone. One report had accurately noted the presence of twenty U-boats in the area; how could the British possibly have known the exact number from direction-finding fixes? Still another Allied signal had reported that two U-boats were at 31° N, 39° W for a possible rendezvous with a supply ship. That was fine: except that the rendezvous, ordered by a radio signal sent to the boats by BdU, had not yet taken place.

Incredibly, the investigators shrugged off all the evidence that their enemies were reading the Enigma traffic. The clinching argument, in their view, was that if the British possessed the cryptological genius required to penetrate the complexities of the Enigma, they would surely not have made such a simple-minded mistake as to use an insecure code like the Naval Cypher No. 3 for their own vital communications.[55]

Meanwhile, everyone in the U-boat command structure with access to operational orders and signals was interrogated and placed under surveillance. Only Dönitz and his chief of staff, Kapitän zur See Eberhard Godt, were considered exempt from suspicion. When this avenue also failed to turn up any source of the leaks, Godt made a small joke. "Shall I investigate you," he asked his commander-in-chief, "or will you investigate me?"[56]

TWO SQUADRONS OF U.S. Army Air Forces B-24 Liberators at Langley Field, Virginia, were the first Allied submarine-hunting aircraft to receive the new centimeter-wave radars, and on November 6, 1942, the first three aircraft of the group took off from Gander, Newfoundland, for the difficult transatlantic crossing. Flying across the ocean was still an intrepid undertaking in 1942. There were no radio beacons or reliable weather data over much of the Atlantic; like mariners in the age of sail, aircraft navigators relied on celestial fixes with sextants to establish their location and more than a little seat-of-the-pants experience to avoid dangerous weather. Over the next several weeks two of the Liberators from the 1st and 2nd Antisubmarine Squadrons vanished while making the attempt to cross the Atlantic and were never heard from again.

Adding to the perils of weather was the usual comedy-of-the-absurd snafus of military bureaucracy. The squadrons were originally intended to join the Allied forces that had landed in North Africa that month, and to operate out of bases in French Morocco. The decision was then made instead to attach them, at least temporarily, to Coastal Command in England while they gained experience operating their new equipment patrolling the Bay of Biscay. The change in plans was news to everyone but the aircrews. An advance echelon of ground and maintenance staff departed from Langley on C-54 transports at the end of November and made it as far as Accra, on the west coast of Africa, where they sat for a month waiting for transport the rest of the way. No one had apparently decided where exactly in England the squadrons were to be based or who they reported to, so several of the Liberators that followed on the southern route via Africa arrived at airfields there with no idea where to go next. They were "passed from station to station in Africa because no one knew where to order them," reported their commanding officer. In mid-January Lieutenant Colonel Jack Roberts was still trying to find four of his planes; he reported to Army Air Forces Antisubmarine Command back at Langley that "they were known to be somewhere in Africa."

Roberts was able to get somebody to decide that they would be based with Coastal Command 502 Squadron at St. Eval in Cornwall, but though the RAF greeted them warmly, no preparations had been made and conditions were frankly abysmal. The field's facilities were already overstretched and there were no hangars available at all for the American planes. Maintenance had to be done in the open air during the limited daylight hours. None of Roberts's men had been issued winter clothing: another administrative slipup apparently due to word not having caught up with their change in orders from North Africa. Roberts spent two days trying to get through by phone to the only American air force authorities he could find in the country, VIII Bomber Command headquarters outside of London. He finally was able to get its commander, Brigadier General Ira Eaker, to issue an order placing his unit under Eaker's command for administration and supply and under RAF Coastal Command for operational control.[57]

Worst of all was the food. The Americans were literally incredulous at how bad the meals were in the RAF mess. "Unbelievably bad," recalled one

American officer. "The mess seemed to be a continuous diet of cabbage. Brussels sprouts were considered a rare treat and once we were given cauliflower with a cream sauce for dessert!" That apparently was the norm in the wartime service; an RAF officer at another base in England recalled endless breakfasts of pig's liver and the mess hall's "combined tea and cocoa urn" that produced tea-flavored cocoa at breakfast and cocoa-flavored tea at supper. Roberts's men were eagerly awaiting the establishment of their own mess, which they had been promised would begin operating in February, and the chance to have "some real American chow."[58]

Phil Morse and Bill Shockley made their own arduous transatlantic crossing a week after the first of Roberts's B-24s. It would be the American operational researchers' first chance to confer with their British colleagues, a step Morse felt was long overdue. He was well aware how isolated the American scientists in general were, and how much they still had to learn from their more experienced British counterparts, who had been fighting a war in earnest for a good two and a half years longer than any of them. "Civilian technical experts here in America tend to stick to their own laboratories and depend on chance contacts from England," he noted upon his return from England early in the new year. "This is a most dangerous practice, which may lead to serious consequences. It is necessary for our technical men to get across as often as possible to see what is going on."[59]

It would also give them a chance to be at the scene where the action was now shifting, with the withdrawal of the U-boats from the American coast, and to play a key part in managing the introduction of the new centimeter-wave radar. Shockley already knew Roberts, having visited Langley several times; both Morse and Shockley were keenly aware of the potential of the new radar in the war against the U-boats if all of the inevitable kinks could be ironed out in its introduction to operational use.

Their trip over was by a long and roundabout route. Boarding a Pan Am flying boat in New York, they flew first to Bermuda, then to the Azores, then to Lisbon, where they arrived in the early evening, the plane circling the harbor for half an hour waiting for the fishing boats to clear a path for them in the harbor. A little after midnight they continued on to Ireland, touching down at the seaplane port in Foyne on the west coast just after dawn, then transferring to a flight departing from nearby Shannon Airport at noon for Bristol. They arrived in London that night. It was Thursday, November 19, three full days after they had left New York. In the still blacked-out city they were driven to their appointments over the next few weeks by "cheerful, iron-nerved girls in uniform" who sped through the dark and often fog-obscured streets.

The following Monday the Americans met Blackett at the headquarters of the U.S. Eighth Air Force, which had been established at High Wycombe, about thirty miles northwest of London. Blackett was "even leaner and quieter than when I had known him in 1931," Morse thought. Morse and Shockley spent the ensuing days steeped in conferences with Blackett, Williams, Gordon, Whitehead, Baughan, and other members of the British operational research groups at the Admiralty and Coastal Command headquarters and visiting an asdic training center in Scotland and a Coastal Command group in Plymouth. Morse did not see Tizard but he did have a session with Sir Stafford Cripps, the new minister for aircraft production who effectively headed the Anti-U-boat Warfare Committee, and proposed that an American representative be added to the committee.

Morse also took every opportunity to press for having some of the British civilian scientists come "to the States to give us help." Blackett on more than one occasion mentioned his concerns about the allocation of the heavy bomber force in the antisubmarine fight. In the discussion at VIII Bomber Command, Blackett noted that the entire premise of the B-17 Flying Fortresses, which were the backbone of the growing American bombardment force in England, was that they would be able to carry out high-altitude, precision daylight attacks. But most of Europe was so frequently blanketed in low clouds that this was unlikely to be possible very often in practice. A better use for the planes, Blackett suggested, might be to carry out ASV radar attacks on submarines in the bay.[60] Morse also noted that while virtually all of the American officers at the bomber commands were convinced that the best way to deal with the submarine menace was to bomb the German U-boat bases along the French coast, "Prof. Blackett, who is the only person I met having quantitative data concerning both operations, is of the opinion that it would be better to attack the submarine after it had left the base." Blackett's preliminary figures that he showed Morse suggested that every 100 hours of ASV patrolling in the bay destroyed three times as many U-boats as the same amount of time spent on missions to bomb the submarine pens. The main problem was that the heavy concrete roofs of the pens were proving virtually impervious to bombs. ("Concrete is quite cheap," Morse noted.)[61]

On December 2, Morse had a long talk with Commander Winn of the Operational Intelligence Centre, and Winn suggested he might find it interesting to come and visit the submarine plotting room, if the required high-level permission could be obtained. A week later Morse was again at the Admiralty and mentioned it to Blackett, who, to Morse's amazement, arranged the visit "in about three minutes." Winn pointed out to Morse on the plot a convoy currently under attack; the Americans, he complained, were still reluctant to reroute convoys and had ignored his plea of a few days earlier to do so, which might have saved the convoy. "Possibly they do not yet believe the plot predictions!" Morse recorded in his diary.

Shockley had meanwhile left for a visit to St. Eval on November 29 and would later return there and end up staying on for an additional three months to help Roberts's squadrons learn to use their radars and refine operational procedures. Riding along on air patrols over the coast of Virginia was one thing, but Shockley was now in a war zone and just how a civilian scientist might be treated if a military plane he was riding in went down and he was captured by the Germans had already raised some qualms. Admiral Furer of the U.S. Navy's R&D department had noted that under international law, an enemy national civilian aboard a naval aircraft had the same status as on a ship at sea; he could be held as a prisoner of war—but not shot as a spy. Furer added, however, that the navy "has a mild doubt that the letter of International Law will be observed" by the Germans. The legal situation was murkier with respect to a civilian on an army plane. In any event, the RAF had adopted the precaution of issuing civilian personnel engaged on scientific missions that took them into a war zone a temporary "honorary commission" as an officer, which allowed them to wear a uniform and presumably be treated as a normal combatant if they fell into enemy hands.[62]

Morse sailed home on the *Queen Elizabeth* on December 21 (the available planes were all booked full of high-ranking officers flying home for Christmas). The luxury cruise liner had been pressed into service as a troop transport and Morse was assigned the bottom tier of a triple-decker bunk hastily assembled of two-by-fours "brutally nailed" into the beautiful wood paneling of what had been a first-class cabin.[63]

A couple of weeks after arriving back in Washington, Morse heard from Shockley of a jarring but revealing incident that had happened at St. Eval while Shockley was there. It was a sobering allegory of the critical importance of the work of the operational research groups in the deadly business of war. One of the Liberator crews had successfully sighted a surfaced U-boat, but when they tried to drop their depth bombs the shackles holding the bombs froze up and would not release. On their return to base they discovered that the shackles had rusted tight in the damp Cornwall air. On their next patrol, January 22, 1943, the crew ran into a pea soup fog; trying to find their way back to St. Eval over the trackless sea at zero visibility, the plane slammed into the Cornwall cliffs, killing all aboard.[64] Morse was prompted to look at the statistics on flying hours per successful attack, and realized with a certain shock that even if the crew of Aircraft S of 2nd Squadron had not suffered that tragic end, their experience was in one sense completely typical: the crew of a sub-hunting aircraft on average encountered a *single* chance to attack a U-boat during all of its operational patrols. "There was no such thing as an experienced antisubmarine air crew," Morse wrote:

> Fighter pilots and ground troops usually fought several battles during their active life; if they were lucky enough during the first few "lessons," they learned how to fight by fighting. But if ASW fliers didn't do things right the first time, they usually had no other chance. . . . Therefore the *only* way to determine the best antisubmarine tactics was for a group such as ours to study *all* the reports of battles with U-boats, made by *all* the crews, and then to determine . . . what to do and what not to do.[65]

It was, Morse realized, the strongest argument yet for the importance of their work. It was also a disquieting reminder of all that was riding on it.

10

A Very Scientific Victory

AT THE FIRST MEETING of the Cabinet Anti-U-boat Warfare Committee back in November 1942, Cherwell was once again at the prime minister's side and a thorn in Blackett's. Cherwell as usual commandeered the technical discussion, peremptorily demanding a study on the effectiveness of air and surface escorts in protecting merchant ships from U-boat attack. The committee assigned Blackett the job of carrying out Cherwell's bidding and was told to prepare "as soon as possible, an agreed statistical analysis" of the data. By "agreed," Churchill made clear, he meant that Blackett was to submit his numbers to Cherwell for approval before forwarding his findings to the full committee.

As irritating as Cherwell's intervention was, Blackett conceded it was the right question to ask. Earlier in the year, Tizard had acknowledged that for all of his run-ins with Cherwell, the very fact that Churchill cared about science was not something to be taken for granted: "What previous prime minister of England ever had a scientific adviser continually at his elbow?" he observed to a parliamentary committee in February 1942. That Churchill listened so much to Cherwell infuriated Tizard, Blackett, and other scientists working on the war. But they were well aware, at least at some level, that it was the price to be paid for having a prime minister who would listen to them as well.[1]

Cherwell's question prompted Blackett to reexamine a fundamental matter that had been largely overlooked, namely just how many escorts were needed for a convoy of a given size, and indeed what the optimal size for a convoy was in the first place. Blackett admitted he had dropped the ball:

> Looking back, I think we operational research workers at the Admiralty made a bad mistake in not realising as soon as the group was formed in the spring of 1942 the vital importance of working out a theory of the best size for a convoy. However, it was not until late autumn that the problem became focused in our minds, largely through discussions that took place at the Prime Minister's fortnightly U-boat meetings. The problem arose as to what was the best division of our limited shipbuilding resources between merchant ships and the anti-U-boat escort vessels. Every merchant vessel completed brought the United Kingdom additional much needed goods; every escort vessel completed added to the protection of the convoys and so reduced their losses by U-boat attacks and so *saved* more ships and cargoes. To make a quantitative comparison of the relative advantage of building escort vessels or merchant ships, one needed to know how many merchant vessels would be saved, that is not sunk, by each extra escort vessel protecting the convoys.[2]

Reviewing statistics for 1941 and 1942, Blackett's team concluded that each additional escort vessel saved approximately two to three merchant vessels a year. That meant that, all things being equal, building one new destroyer a year was worth the same as building two or three new merchant ships a year. But, as Blackett also discovered, that was a finding of more theoretical than practical value, as shipyards could not easily change from producing one type of vessel to another: they were already pretty much committed to the building programs already under way.

The point about convoy size, though, leapt out from the analysis. It also plunged Blackett at once into yet another controversy with the military establishment. The Admiralty had long maintained that smaller convoys were preferable. Forty ships was considered the best size; convoys of more than 60 ships were flatly prohibited. There was no dispute that large convoys were harder to handle and manage and tended to overload port facilities. The navy, however, went further, insisting that small convoys also afforded the best protection.

The statistics disagreed. Blackett's group, in an analysis completed January 27, 1943, found that each ship in a small convoy of 15 to 24 ships had a 2.3 percent chance of being sunk when the convoy was attacked by a

U-boat, versus only 1.1 percent in a large convoy of 45 or more ships.[3] The British navy had an obscure rule for establishing how many escort vessels were required for convoys of varying size. It probably went back to the First World War; Blackett was never able to track down exactly where it had come from. It gave the number of escorts as $3 + N/10$, where N was the number of merchant ships in the convoy. Blackett pointed out that the rule even on its face contradicted the navy's assertion that small convoys were safer. By this formula a 20-ship convoy required 5 escorts, while a 60-ship convoy required 9 to provide a commensurate level of protection. Yet if that were the case, why not combine three 20-ship convoys and allocate all 15 of their escorts to the resulting single 60-ship convoy, thereby almost doubling the number of escorts over that specified by the Admiralty's formula?[4]

On February 5, Blackett presented his preliminary report on the value of escorts and aircraft to the Anti-U-boat Committee. (A cover note stated, "The paper has been submitted and discussed in detail with Lord Cherwell during the various stages of preparation. The responsibility for facts and figures rests however with Professor Blackett.") The paper reaffirmed the value of air cover, especially VLR aircraft; even when aircraft were unable to carry out a successful attack, their mere presence forced the U-boats to submerge, and then often lose contact with the convoy. An average of even 8 hours of flying a day by VLR aircraft in support of a threatened convoy reduced losses by 64 percent. In its average service life of 40 sorties, each VLR aircraft saved 13 ships that would otherwise be torpedoed. But a key recommendation of Blackett's was to increase at once the size of convoys. In fact, he pointed out, the number of ships lost in a 60-ship convoy was on average almost exactly the same as the number lost in a 20-ship convoy.[5]

Several more weeks of analysis and discussion among Blackett's group produced some further insights into what was going on. A convoy's size made almost no difference in the chances that the convoy would be sighted and attacked by a U-boat. It also made almost no difference in the absolute number of ships that would be sunk once a U-boat had penetrated the destroyer screen around the convoy, since there were always more than enough targets and each U-boat had a limited number of torpedoes. So there was safety in numbers simply because one ship was less likely to be picked out from the crowd. Looking at the actual number of ships torpedoed per convoy sailed (including stragglers who failed to keep up with the escorts), the researchers found that it worked out to almost exactly 0.9 ship per convoy regardless of convoy size.[6]

The safety of large convoys

Patrick Blackett and J. H. C. Whitehead showed that doubling the number of ships in a convoy increased by only ⅙ the number of escort vessels needed to maintain the same protective cordon around its perimeter.

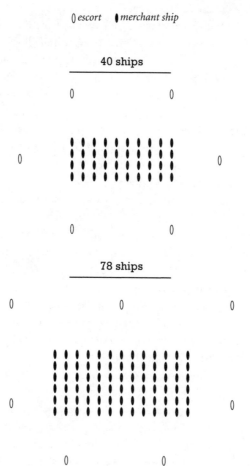

〇 *escort* ❙ *merchant ship*

40 ships

78 ships

The mathematician J. H. C. Whitehead shortly afterward produced a paper explaining further why a large convoy did not need as many escorts per ship to afford the same level of protection. Even the $3 + N/10$ formula underestimated the benefits of combining small convoys. The chances of a U-boat penetrating the escort screen, Whitehead noted, was basically a function of the linear spacing of the escorts around the convoy's perimeter. But the length of the perimeter increased very slowly with increasing size of the convoy. It was a familiar geometric principle that occurred frequently in biology: the area within a perimeter increases with the square of the perimeter's length. To maintain the same linear spacing around the perimeter, the number of destroyers required increased as the square root of the number of ships they were assigned to guard. A 78-ship convoy had a perimeter only one sixth longer than that of a 40-ship convoy; 7 escorts on the larger convoy provided the same level of protection as 6 escorts on the smaller one. Thus by far the most efficient use of the available surface escorts was to employ them in the largest possible convoys practical. Increasing the average size of convoy even by 50 percent, Blackett calculated, would reduce the total number of merchant ship sinkings by half.[7]

Blackett was well aware what a serious matter he was taking on his shoulders. He and his co-workers "had proved intellectually to themselves that big convoys were safer than small ones," as Blackett later said, but "before we advised the Navy to make this major change, we had to decide whether we really believed in our own analysis." Blackett did so for himself by deciding "that if I were to send my children across the Atlantic at the height of the U-boat attacks I would have sent them in a big rather than a small convoy."[8]

But at the Anti-U-boat Committee the Admiralty was unimpressed, and dug in its heels. At the February 24 meeting, Lord Leathers, the minister of war transport, urged that the artificial limit of 60 ships be waived as the only way to relieve the increasing backlog of supplies awaiting shipment across the Atlantic. The first sea lord, Admiral of the Fleet Sir Dudley Pound, replied that the "sailing of such large convoys was asking for trouble" and reiterated that 40 ships was the optimal size.

Blackett tried again a week later, pointing out that even if losses increased with larger convoys, total imports would increase as well given that significant numbers of ships were now being delayed by the 60-ship rule:

More ships are now presenting themselves for convoy than can be accepted within the 60 ship maximum rule. The reason for the imposition of this limit is the expectation of heavy losses with larger convoys. The object of this note is to point out that it is inconceivable that the increase of losses could be large enough to offset the immediate gains in imports by adopting larger convoys.[9]

But, he pointed out, the assumption that losses would increase at all was "unduly pessimistic"; he was just presenting a worst-case possibility, and the evidence was that losses would likely not increase even in absolute numbers. Blackett advanced all of these arguments yet again in an unusually forceful paper produced later that spring (forthrightly titled "The Case for Large Convoys"). In it he also noted that a related advantage of fewer sailings with larger convoys was that the available long-range and VLR aircraft would be able to cover a greater percentage of ships during their perilous passage through the still inadequately protected "air gap." The calculation was rather complicated, but conferring with the Coastal Command ORS he predicted that air cover would increase 40 percent.

In the end it was the press of import requirements that forced the Admiralty to back down. Blackett's arguments helped offer some assurance that the change would at least not be a disaster, even if he had failed to convince Pound that it would be a virtue in itself. By late spring and early summer 1943, 80-ship convoys were common. As Blackett had predicted, the large convoys were far more economical in their use of surface escorts. The change allowed a large number of destroyers to be freed up the following year to support the D-Day landings. Blackett later calculated that had he thought to look into the problem in spring 1942 instead of a year later, a million tons of shipping, 20 percent of the total losses, could have been saved in the interim through the adoption of large convoys. But in the rush of other work no one in the Admiralty operational research section had thought of it.

Still, no one on the navy staff had thought of it, either. "The problem forced itself on our notice," said Blackett, only when the operational researchers began exploring the tangential question Cherwell had raised about the relative effectiveness of escorts. "As in most of the important cases . . . the really vital problems were found by the operational research groups themselves rather than given them to solve by the Service operational staffs."[10]

THE U.S. EIGHTH AIR FORCE'S ENTHUSIASM for bombing the German submarine bases on the Bay of Biscay, which Morse had encountered during his visit to England in the fall of 1942, was the bastard offspring of mixed motives.

The rapid buildup of the American B-17 and B-24 forces in England since the summer had exposed glaring deficiencies in the organization, training, and abilities of the bomber units. There was no lack of enthusiastic volunteers: the Army Air Forces had so many recruits that by the end of 1942 it had a backlog of 100,000 men awaiting flight training and even enlisted technical positions were so oversubscribed that air force officers estimated they already had all the candidates they needed for the next two years of all-out expansion. In December 1942 voluntary enlistment in the AAF was ended.[11] "The Romance of the Air," remarked the novelist E. M. Forster in a letter he wrote his friend and fellow writer Christopher Isherwood, "is war's last beauty parlor."[12]

The problem facing the AAF was how to train all of those eager young men fast enough to get them into combat units. Its solution was a kind of assembly-line education that seemed to teach almost nothing. Demoralized instructors, most of them civilian teachers promised rank, promotion, and the chance to put their talents to use serving their country, found themselves wearing private's stripes and delivering canned lectures they were ordered to follow by rote. One student pilot remembered the eight classes a day he attended as "the saddest, poorest, most incomplete" attempt at education he ever experienced in his life, "the maximum of predigested information in the minimum time."[13]

Deployed to England, the crews were ill-prepared and their aircraft plagued with mechanical problems. An inspection report found radio operators who could not send and receive Morse code at the minimum rate of eighteen words per minute, gunners unfamiliar with how to operate their turrets or track a rapidly closing fighter, pilots who had scant experience flying at high altitudes, bombardiers who could not read maps. Most of all, commanders concluded, the men just needed some toughening up, which could only come from actual combat experience of flying through flak and enemy fighters, enduring the freezing cold and low oxygen of high altitude in unheated and unpressurized cabins. From this perspective, bombing the Bay of Biscay ports, which were much closer than targets in Germany, offered an opportunity for the American crews to get their feet wet.[14]

Meanwhile, criticisms in the American press about the slow pace of the Eighth Air Force's buildup were making the U.S. air force commanders and their political masters in Washington worried. In November 1942,

H. H. "Hap" Arnold, the commanding general of the U.S. Army Air Forces, told General Eaker he was sending from Washington an officer especially qualified for the task of "writing up and presenting to the American public" the success of the heavy bombers to date and explaining their ability to "crush Germany's capacity to wage war at its source." A few months later the assistant secretary of war for air, Robert Lovett, nervously called Eaker's attention to a recent article in *Reader's Digest* questioning why the American strategic bombers were not yet doing the great things that had been promised of them.[15] Lovett and Arnold were both eager for some stories that would offer quick proof of the value of U.S. strategic air power to the war effort, and deflect pressures to divert aircraft to other uses or to other theaters such as North Africa, where the airmen would not have the same opportunity to achieve the decisive victory through air power—alone— that they so fervently believed in.

In truth, neither the Eighth Air Force nor the RAF commanders were ever that enthusiastic, deep down, about the operation against the U-boat pens. But they understood politics, too, and thought it might at least divert attention away from the worse prospect of having more of the heavy bombers diverted to antisubmarine patrols. Like Churchill, Roosevelt was also pressing for action in the fall of 1942 to stem the losses of merchant shipping. Eaker confidently assured the president that with 1,000 bombers he could reduce German submarine operations in the Atlantic by 60 percent by destroying the Bay of Biscay bases. Since he had far fewer than that number he was probably hedging his promise to deliver results, while also sensing an opportunity to get more airplanes. In mid-October, with General Dwight D. Eisenhower calling for the defeat of the U-boats as "one of the basic requirements to the winning of the war," Eaker again pointed to bombing the U-boat bases as the best way to get the job done. "Given enough bombers here we can destroy the submarines where they are built and launched more economically than looking for a needle in a haystack—a submarine in the Atlantic."[16]

On October 21, Eaker carried out the first raid, sending 90 B-24s and B-17s against L'Orient. Only 15 planes reached the target, the rest turning back because of bad weather. Over the next ten weeks the Eighth Air Force carried out nine more attacks on the U-boat ports, dropping 1,500 tons of bombs and losing 28 aircraft and 300 crewmen killed, wounded, or missing in the process. Analysts examining the attack photographs found that only

6 percent of the bombs dropped by the American aircrews fell in the "target area," and even that was being generous, since, as they acknowledged, it was based on "a very liberal interpretation of the word 'Target,' and it includes all hits and craters which can be seen."[17]

The AAF's own scientific experts had pointed out from the start that the concrete roofs of the submarine pens were virtually indestructible and that the related services that supported the port operations were widely spread out in the French towns, so that bombs which missed their target would fall on little of any consequence. "Based upon experimental data, involving tests conducted during the past year in the United States, it is my opinion that none of the U.S. bombs now available to this Command are capable of perforating the roofs of these pens, at least from any practicable bombing height," the VIII Bomber Command's operations research section noted on December 8.[18] A report by the British Ministry of Economic Warfare reached the same conclusion earlier. A memorandum by the British Air Staff in November also noted that the amount of fuel, water, electricity, and general supplies required by the bases was such a small percentage of the total available in each of the port cities that even widespread destruction of the towns themselves would have little effect on U-boat operations from their ports.[19]

Hoping for better results, U.S. air commanders ordered the 31 B-17s assigned to a raid on Saint-Nazaire on November 9 to fly as low as 7,500 feet to improve bombing accuracy. It was practically suicidal. Three planes were shot down and 22 damaged by flak. Once again the photographic interpreters strained to find any evidence that their bombs had done any damage at all. Only 75 of the 344 bombs dropped even showed up on strike and reconnaissance photos of the entire target area; of those, only 8 fell within 600 feet of the port facilities or workshops the planes had been assigned to hit. Saint-Nazaire was hit by five heavy raids in early November; according to agents' reports the port was back in full operation two weeks later.[20]

Harris was nothing if not consistent: the head of Bomber Command vehemently opposed attacking even the U-boat bases as a diversion of effort from the strategic bombing campaign against the German heartland. But the British Air Staff was willing to apply to the U-boat ports the methods of incendiary-fueled area bombing that Harris had made his trademark. With both First Lord of the Admiralty A. V. Alexander and Churchill now strongly backing the proposal, the War Cabinet on January 11, 1943, agreed that Bomber Command should make as its first priority the destruction of "the whole area" around the U-boat pens. The RAF raids began the night of January 14 with a strike on L'Orient.

The effort received further endorsement at the highest levels on January 21, when Roosevelt and Churchill, meeting in Casablanca, issued a joint directive for the Allied bomber offensive. The agreement still emphasized that the prime objective of the British-American strategic bombing campaign was "the progressive destruction and dislocation of the German military, industrial, and economic system, and the undermining of the morale of the German people to a point where their capacity for armed resistance is fatally weakened"—which meant keeping the focus of the bomber campaign on striking the German homeland. Eaker, who had a degree in journalism, had written a clever one-page pitch that basically allowed the American and British air forces to keep doing what each wanted to do anyway but presented it as coherent grand strategy. Harris's night area bombing and the AAF's daylight "precision" bombing were hardly the same thing, but Eaker hit on a phrase that Churchill seized on at once when he handed him his paper at Casablanca: "If the RAF continues night bombing and we bomb by day, we shall bomb them round the clock and the devil shall get no rest." Churchill loved the bit about "the devil" and started using the phrase himself.[21]

The Casablanca directive did also, however, note that besides the priority objectives in Germany there were other targets "of great importance from the political or military" point of view that deserved attention, the U-boat bases in France being the prime example.[22] Seven more fire-bombing raids on the French ports followed over the next four weeks. Some parts of L'Orient were hit with one incendiary bomb per square yard. Though civilian casualties were surprisingly light—more than 20,000 of the city's 40,000 residents evacuated the city after the BBC broadcast a warning to the population in January—fires set by the air raids burned 3,500 of the town's 5,000 buildings to the ground, all major utilities were knocked out, and almost all of the remaining civilian population fled.

Raids against Saint-Nazaire followed through the spring, with 1,500 tons of incendiaries and 300 tons of high explosives. The town was destroyed in huge blazes that consumed every major building along with workers' housing, schools, churches, and hospitals. Analysts poring over reconnaissance photos finally managed to identify a dozen direct hits on the submarine pens: the bombs had left barely visible pockmarks in the concrete. Dönitz summarized the effects of the raids with the contempt they deserved:

> The Anglo-Saxons' attempt to strike down the submarine war was undertaken with all the means available to them. You know that the towns of Saint-Nazaire and L'Orient have been rubbed out as main submarine bases. No dog or cat is left in these towns. But the U-boat shelters remain.

Though the towns were rendered virtually uninhabitable, rail lines and power plants damaged in the raids were quickly repaired. Some critical repair shops and port services were destroyed, but rather than abandoning the bases, Dönitz simply moved those operations into the protective shelter of the concrete pens themselves. A year and a half later even the air commanders abandoned all remaining pretense that the raids had accomplished anything. The attacks had "caused inconvenience," a U.S. air staff intelligence report concluded in December 1944, "but have never in the long run affected the operational use of the bases."[23]

The submarine shelters themselves have defied all subsequent attempts to demolish them: they remain to this day, useless, abandoned, indestructible witnesses to the thoroughness of the Thousand Year Reich and the futility of strategic bombardment as a shortcut to military victories.

THE CIRCULAR LOGIC of the bomber barons was breathtaking. As Harris saw it, the only things even worth destroying from a military standpoint were the things that his bombers could destroy; if he could not target it, it was not a target in the first place. Even after the Casablanca directive Harris continued to insist that the U-boats were a waste of time. He was either sure enough of his ability to reverse Churchill's orders or so tone deaf to the politics that he continued to issue forth streams of sarcastic and hyperbolic memoranda disparaging any notion that heavy bombers ought to be used for anything but bombing Germany. The Admiralty's proposals to bomb the submarine pens and increase the number of sub-hunting VLR aircraft with centimeter-wave ASV radar would leave Bomber Command a "residuary legatee," left to pick up the scraps to carry out its prime mission after "all the other claimants press for their full, real or fancied, requirements being met," he furiously insisted in a March 29, 1943, memorandum for the

Anti-U-boat Committee. "The employment of aircraft to attack the fringes rather than the centre of the objective is a highly extravagant process," he continued:

> In view of the very large number of U-boats which the enemy will oper-
> ate in the coming months, the proportion of his successes which would
> be eliminated by accepting the Admiralty proposals seems to me to be
> so small as to be negligible. The effect of them on the Bomber Offensive
> would be catastrophic. . . . In the present case it is inevitable that at no dis-
> tant date the Admiralty will recognise that U-boats can be effectively dealt
> with only by attacking the sources of their manufacture but by then much
> time will have been lost and the whole success of the Bomber Offensive,
> which may have a decisive influence on the success of Russia and even of
> her remaining in the war, will have been jeopardised. . . . This in my opin-
> ion would be a far greater disaster than the sinking of a few extra merchant
> ships each week.

Harris dismissed as "purely defensive" the idea of "chasing wild geese on the Bay of Biscay" or otherwise employing long-range aircraft on "seagoing defensive duties." By contrast, the bomber offensive was "the only effective means open to the United Nations in the immediate future for striking directly at Germany." Indeed, Harris insisted, it was already close to winning the war: "Opportunities do not last for ever, and we have got so near with the existing bomber force to producing a state of destruction and chaos in Germany insupportable to the enemy, that to let up on it now would give him new encouragement, and would make it very difficult, if not impossible, to catch up again."

Bomber Command also was in a fight for allocation of centimeter-wave radar sets, which added to Harris's opposition to U-boat hunting. The same technology was used in a radar system being developed for "blind" bombing through clouds, code-named H2S; in principle this would give the bombardier a picture of a coastline or other ground features below to help locate the target under poor visibility. Harris insisted in his memorandum that the technology was vital to his air campaign against Germany, but unlikely to make much difference to the anti-U-boat effort—incredibly revealing in the process that he did not understand the first thing about radar or the U-boat search problem:

I feel, however, that too much emphasis is being given to the possibility of locating U-boats by means of A.S.V., and too little to the difficulty of attacking them successfully when they are located. Our experience, which is considerable, is that even expert crews find it no easy matter to attack with accuracy even a city by means of H2S. I am therefore rather sceptical of the prospects of inexperienced crews with A.S.V. Indeed, I feel that the provision of aircraft equipped with this apparatus will mark the beginning rather than the end of the difficulties involved in sinking U-boats.[24]

Blackett penciled "nonsense!" next to that paragraph on his copy. In fact, for both technical and military reasons centimeter-wave radar worked much better as an antisubmarine weapon than as a high-altitude bombing aid. A submarine's metal conning tower generated a strong and clear radar reflection in the centimeter-wave band: trials had confirmed that the new radars could easily pick out a surfaced U-boat at a range of twelve miles. By contrast, the radar echoes coming off large geographic features when the radars were used for ground mapping in the H2S sets were always vague and extremely difficult to interpret.

Moreover, there was the acute danger that an H2S-equipped bomber would sooner or later be shot down over enemy territory, giving away to the Germans the secret of the new Allied radar; once that happened it would only be a matter of time before they developed countermeasures, including a warning receiver that could alert U-boat crews of an approaching aircraft operating a radar at this new shorter wavelength. That was indeed exactly what did happen. During the second bombing mission flown with the new device, February 2, 1943, a bomber carrying an H2S set was shot down over Rotterdam. A complete report on the new technology reached the highest levels of the German command a few weeks later; work began at once on a centimeter-wave U-boat warning receiver, code-named Naxos. Only some fortuitous technical glitches prevented its being ready in time to thwart the Allies' major offensive against the U-boats in the Bay of Biscay a few months later, in the summer of 1943.[25]

IT WAS REMARKABLE that to Harris, bombing submarine factories in Germany was the only "direct" means of attacking the U-boats while dropping depth charges on them when they were hunted down at sea was going after "the fringes." But a year later he would be making similar arguments to oppose the Normandy landings, insisting that the bomber offensive offered the only sure route to victory, and characterizing the imminent seaborne and land offensive against Germany as risky, even unnecessary. (General Alan Brooke, the chief of the Imperial General Staff, noted dryly in his diary after one meeting with Eisenhower and his top commanders preparing for D-Day, "Harris told us how well he might have won the war had it not been for the handicap imposed by the existence of the other two services.")[26]

The Admiralty and the Air Staff exchanged increasingly vituperative notes over the bombing of the U-boat bases. Blackett noted in one that if the price to be paid for Harris's bomber offensive was to delay by six months the launching of a ground campaign in Europe because of a shortage of transport needed to carry troops and supplies across the Atlantic, then "the word 'offensive' is as wide of the mark as the word 'defensive' when applied to the bombing of U-boat bases."[27]

Harris, true to his convictions and his disdain for "panacea bombing," did not hesitate to join sides with critics who were doubting that the air force was achieving much of anything by attacking L'Orient and Saint-Nazaire. That put him at odds with the U.S. air force as well as with his fellow RAF officers who were directing the operation against the U-boat ports.[28] Yet few air senior officers took issue with Harris's larger point, that the strategic bomber offensive was the core mission and raison d'être of the air forces. Blackett's involvement in the argument reawakened the previous year's dispute over bombing policy, and RAF officers closed ranks to defend their service. Blackett probably did not help matters by going toe-to-toe with Harris in an exchange of acerbic memoranda about fundamental war strategy; he began his reply to Harris's March 29 memorandum, "I have read the paper by A.O.C. in C., Bomber Command, which is coloured by the same fundamentally false strategic conception displayed all through these discussions with the Air Staff."

Even the new commander of Coastal Command, Air Marshal John Slessor, bridled when Blackett and Williams recommended to the Anti-U-Boat Committee the transfer of VLR aircraft from Bomber Command to increase patrols in the bay. Blackett had not bothered to discuss their paper with Slessor before presenting it to the committee, an amateurish political miscalculation. Slessor, offended, reacted by dismissing the scientists' calculations of the effectiveness of patrolling the bay as "slide-rule strategy of the worst kind"—and later insinuated that Blackett's real motive was that he "was intellectually and temperamentally opposed to the bomber offensive." In his memoirs Slessor tried to justify his stance by suggesting that Blackett and Williams were simply out of their depth:

> The operational research scientists had no stronger supporter than I, and these two were among the best of them. But they must stick to their lasts. Statistics are invaluable in war if they are properly used—in fact you can't fight a modern war without them. But the Bay offensive was a battle, and a bitterly contested one, and nothing could be more dangerously misleading than to imagine that you can forecast the result of a battle or decide the weapons necessary to use in it, by doing sums. It is not aeroplanes or ships or tanks that win battles; it is the men in them and the men who command them. The most important factors in any battle are the human factors of leadership, morale, courage and skill, which cannot be reduced to any mathematical formula.[29]

In other words, calculations based on quantifiable data were invaluable—but a commander was always free to ignore them on the grounds that unmeasurable factors were still more important. The fact that Blackett's and Williams's "sums" had repeatedly tripled or quintupled the effectiveness of the men and their weapons apparently went right by the air marshal, as did the fact that nothing is more bracing for men's morale than success.

Blackett was excluded from a meeting of the Anti-U-boat Committee on March 31 on the slim pretext that there were too many attendees. Picking up on Slessor's line about "slide rule strategy," Air Chief Marshal Charles Portal, the chief of the Air Staff, told the committee he was reluctant to transfer any aircraft from the bomber offensive to the Battle of the Atlantic based "on a theoretical calculation." Harris added his familiar scorn and Cherwell backed him up with a sage pronouncement, based on no apparent facts whatsoever, that adding a modest number of aircraft to the bay offensive was unlikely to be effective. Outnumbered and outargued, the Admiralty representatives were forced to accept Portal's token offer that "inexperienced" bomber crews could be used to continue the raids on the bay ports, as could the U.S. Eighth Air Force on days when clouds obscured the targets in Germany that the Americans hoped to strike with their daylight precision tactics.[30]

THE MATTER, however, was being overtaken by events in Washington, and the mid-Atlantic.

However much Harris, Arnold, Eaker, or other airmen might entertain visions of victory through aerial bombardment alone—and however much they complained about chasing geese and searching for needles—that was not a view that enjoyed much sympathy among the men busily planning for the greatest landing operation in the history of war. The U.S. Army high command was becoming increasingly alarmed over what the U-boat threat was doing to its plans for D-Day. On March 1, 1943, a conference in Washington was convened that brought together a host of British, American, and Canadian sea and air officers to discuss the convoy situation. Admiral Percy Noble, the chief of the British naval staff in Washington, headed the British delegation to the Atlantic Convoy Conference and he noted in his opening remarks, "The submarine menace, to my mind, is becoming every day more and more of an air problem."[31]

Edward L. Bowles, an MIT electrical engineer who had become "Expert Consultant to the Secretary of War" on radar—and, increasingly, the main scientific adviser on antisubmarine air operations by the AAF in general—provided a memorandum to General Marshall on March 3 warning that the ability to launch and supply a land offensive on the continent depended crucially on solving the U-boat problem. Given how much of the total Allied shipping capacity was fully committed just to maintaining Britain's basic necessities such as food and fuel, even a small change in the number of available ships translated into a huge difference in the number of extra men and weapons that could be transported across the Atlantic. Bowles calculated that reducing the rate at which ships were being lost from the current 2.6 percent per month to 1.5 percent would increase by 2 million the number of U.S. troops who could be sent overseas and supplied by the end of 1944. The army's plans for future operations in Europe assumed at a minimum cutting the losses to 2.15 percent.

Bowles, who had drawn on both Blackett's and Morse's work in preparing his memorandum, noted as well that aircraft were ten times as effective as surface ships in sighting a U-boat, and about as effective in sinking a U-boat once it was sighted. The key was to increase sightings. That inescapably meant more aircraft, especially centimeter-wave-radar-equipped VLR aircraft.[32]

The conference had little difficulty agreeing that 260 VLR aircraft were immediately needed for antisubmarine operations, versus the 52 currently available. Making it happen was, as always, another matter. That depended on decisions that were, in the military argot, above the conferees' pay grade. The conference subcommittee charged with "implementation" put it cautiously:

> In view of the shortages of aircraft in all theaters of active operation it would be most undesirable to divert to ASW any *large* number of aircraft now ear-marked for these theaters, but the urgent need for a total of 260 VLR for A/S operations justifies the recommendation that the Combined Chiefs of Staff explore the possibilities of diverting 128 to ASW in addition to the 132 which are to be assigned by 1 June in accordance with present plans.[33]

Bowles was stunned a few weeks later when General Arnold insisted to him that a letter from his own staff the previous summer agreeing to assign even those 132 B-24s to antisubmarine operations "means nothing." The allocation of aircraft, Arnold said, was a decision for the Joint Chiefs, and they had not yet acted on the matter. Bowles rightly suspected that part of the reason they had not acted was that Arnold and the AAF brass had made sure the issue never came up. "To me the shocking aspect of this," Bowles wrote Secretary of War Henry L. Stimson, "is that, since the conference I had with you and General Arnold on August 10, 1942, we have been struggling to obtain an honest-to-goodness allocation of long range aircraft to antisubmarine warfare."[34]

Events at sea, however, were beginning to speak louder than the words of staff officers in Washington. On March 10, the Bletchley code breakers suddenly stopped being able to read the U-boat Enigma signals, the result of the Germans' switch to a new weather short-signal book. Almost simultaneously, the B-dienst broke back into the Allied convoy code, which it had been having trouble cracking since December when a new, more secure procedure was introduced in that system. On March 16, forty U-boats converged in the mid-Atlantic on two eastbound convoys that had sailed from New York the first week of the month. Twenty-two merchantmen were torpedoed and sunk in a single action that cost Dönitz a single U-boat. The tonnage loss was 146,000.

Two days later FDR sent a note to King and Marshall. Referring to the code names for the Allied invasion of Sicily planned for the summer and the buildup of U.S. forces in England in preparation for a cross-Channel invasion of France, the president wrote, "Since the rate of sinking of our merchant ships in the North Atlantic during the past week has increased at a rate that threatens seriously the security of Great Britain, and therefore both 'Husky' and 'Bolero,' it seems evident that every available weapon must be used at once to counter the enemy submarine campaign."[35]

Pressure was also mounting from both the British and the U.S. Army for a total overhaul of the U.S. antisubmarine air effort, much as Captain Baker had urged almost a full year before, to unify the still woefully fragmented command structure. Despite the foot-dragging by Arnold and the other top AAF commanders in allocating additional aircraft, the army had nonetheless taken serious steps to embrace the job as its own. It had renamed the I Bomber Command the Army Air Forces Antisubmarine Command in October 1942 and established a close working relationship with the sea frontier and naval district commands, as well as informal liaisons with the navy's ASWORG scientists. On March 22, 1943, one of the army Liberators at St. Eval sank the group's first U-boat; their success affirmed the army fliers' growing restlessness with what they saw as the U.S. Navy's excessively defensive use of air power to escort convoys.[36]

On April 19, Marshall sent his most strongly worded plea yet for breaking the impasse over the supply of VLR aircraft and giving the whole antisubmarine effort the urgency it needed. His memorandum was addressed to the Joint Chiefs of Staff but it was really aimed at King:

> I am deeply concerned with the matter of the present organization and technique for the employment of aircraft in the anti-submarine effort, and I therefore propose that this be considered by the Joint U.S. Chiefs of Staff in all of its ramifications. . . . I personally am now strongly of the opinion that the ultimate solution for the employment of the Air arm in connection with anti-submarine operations, particularly, and possibly exclusively as applied to VLR aircraft, should be based on an organization under a unified command responsible for temporary allocation and technique of employment. This would result in overriding what appears to me to be the limiting effect of its present employment under the system of naval districts and sea frontiers.

The navy's geographical command system was the nub of the problem, making it nearly impossible to quickly shift aircraft from one area to another. The entire arrangement was built around King's still limited view of aircraft as fundamentally escorts for convoys. Marshall proposed organizing at least

a "highly mobile striking force" of VLR aircraft "for offensive action," which
would be placed directly under the Joint Chiefs in a manner analogous to a
theater command—in other words, taking the unit out of the navy or army
chain of command altogether. He suggested that in addition to the 75 army
and 60 navy VLR B-24s already allocated to antisubmarine operations, each
service would provide an additional 12 per month in May, June, and July to
bring the total up to just over 200.[37]

King was playing several games. He still was adamant about avoiding any joint command that would place a British officer in charge of the
Battle of the Atlantic. At the Convoy Conference, his opening remarks
pointedly vetoed any such suggestion. He said that it was his "very strong
personal opinion" that any command structure with "mixed forces" was
to be avoided. They offered "only the form and not the substance of unified effort," he insisted. King also poured several buckets of cold water on
offensive air operations against U-boats. "I have heard something about
'killer groups,'" he told the conference, and then dismissed them just as
Eaker had, as a search for the "proverbial needle in a haystack." King had
not kept pace with either tactical or technical developments. His justifiable
skepticism about surface "killer groups" simply did not apply to the demonstrated value of radar-guided air search. Bowles, now joined by Vannevar Bush, kept trying to point this out; on April 6 he wrote Stimson about
another of King's pronouncements on the matter, "Dr. Bush and I both feel
that the great weakness of his position lies in his minimizing of the importance of the air attack." From London, where he was currently serving as the
ASWORG liaison to Britain, Arthur Kip weighed in later that spring with
the latest evidence showing that Blackett and Williams's calculations on the
effect of radar-equipped aircraft in the bay had been almost precisely borne
out in preliminary trials. "The point I wish to stress is the extreme importance of acting at once in getting more B-24's over here," Kip wrote, noting
that both the Admiralty and Coastal Command were convinced, too. In
late May, Admiral Stark, the commander of U.S. Naval Forces in Europe,
joined Slessor and Pound in formally requesting the immediate transfer of
seventy-two American B-24s from antisubmarine bases on the east coast of
America to Britain to join the bay offensive. Even Slessor was by now sold
on the idea.[38]

King kept balking and equivocating. For one thing, he had his eyes on
the Pacific, where he was nursing ideas of edging the army out of the long-
range bombardment mission; the navy had been sending much of its allot-

ment of new B-24s to the Pacific theater, equipped not as sub hunters but as bombers.

Only King could afford to stiff the president of the United States. He tended to view FDR as just another meddling civilian, and his resolute policy was to ignore meddling civilians. King routinely refused to see the Navy Department's civilian assistant secretaries ("I didn't have time to educate those people," he disdainfully explained), and when Secretary Knox once sheepishly sent a mutual friend of the two men as an emissary to complain that the admiral never told him anything that was going on in his department, King shot back, "Why should I? The first thing he does is to tell the reporters everything he knows." King felt he knew how to handle the president, too, whom he never warmed to. "Roosevelt was a little tricky," King later remarked, "and in some ways the truth was not always in him."[39]

King was himself a master of the backdoor bureaucratic game, however, and it was becoming clear to him that Marshall's proposal to place the antisubmarine air forces under a single command was going to prevail unless King found a way to get it off the table once and for all. Just when it looked like he would have to give way, King surprised everyone by announcing that he had established a new navy command that would take charge of the entire Battle of the Atlantic.

To give it the maximum authority, King designated the new organization a fleet within the navy command structure—and promptly named himself its commander. The Tenth Fleet had no ships of its own but assumed operational control of all antisubmarine forces in the Atlantic. King then proposed a simple "horse trade," as its critics disparagingly but accurately described it: he would give to the army the navy's pending allocation of B-24s in exchange for the army's existing seventy-seven antisubmarine-equipped versions of the planes, and the army would just get out of the ASW business altogether. It was a bureaucratic and institutional solution rather than a military one and Marshall, who had tried to keep the discussion focused on substance rather than service politics, was clearly dismayed, but accepted it as the only way to break the impasse. The agreement was formally accepted on July 9.[40]

ASWORG moved to the Tenth Fleet with the reorganization. It found a solid champion in the man King had brought in as his deputy, Rear Admiral F. S. "Frog" Low, whom King had recalled from a command in the Pacific to take the job. (The admiral's nickname, Phil Morse noted, "referred to his voice, not his manners.") Low largely accepted ASWORG's views on doctrine, tactics, and even offensive air operations in the bay and incorporated them wholesale into the doctrine and policy papers he issued for the Tenth Fleet, where necessary skirting King's sensibilities simply by employing cryptic or evasive wording (such as calling the bay offensive an attack on "focal areas of concentrations").[41]

ASWORG's role by the summer of 1943 was becoming so routine that it scarcely even raised eyebrows anymore that a bunch of very nonnaval civilians was setting policy on the most minute details of naval operations and doctrine. In December 1942 the group had acquired a roomful of IBM machines and now had a complete set of attack records punched onto IBM cards, which could be sorted and analyzed to answer new questions that arose. The scientists continued to crank out studies assessing such questions as the optimal height of patrol aircraft to fly, the most effective allocation of a plane's payload between fuel and depth bombs, and the right spacing and pattern for depth charges fired by new weapons being developed for ships that could throw a salvo of charges over the bow so that an escort vessel could attack a U-boat ahead of the ship without losing sonar contact. For that last problem George Kimball had hit on an innovative technique of using the IBM equipment. He drew an outline of a submarine onto an IBM card, punched it out, then traced the pattern onto a series of other cards, repositioning it each time according to a table of random distributions. He then had a set of submarine silhouettes that could be "attacked" with different depth charge patterns and the number of hits counted for each. It was an early example of using computing machinery for what was called a Monte Carlo simulation, a useful technique when an analytical solution using mathematical formulas was too complex. Slessor's barbed metaphor was out of date. Strategy by slide rule had become strategy by IBM machine.[42]

EVEN BEFORE the recommendations of the Convoy Conference and the reorganization of the U.S. antisubmarine command structure could be implemented, a remarkable turnaround in fortunes had occurred in the Atlantic. On March 18, 1943, after two weeks of frantic efforts, the Bletchley code breakers resumed reading the U-boat Enigma signals by employing their longest shot yet; piecing together cribs from short-signal messages transmitted by the U-boats to report convoy sightings or their position or fuel supplies, they managed to break back into the system, in many cases by making truly inspired guesses at the contents of the messages to build up a crib to run on the bombes. Signals that reported a U-boat's location might be cribbed by direction-finding fixes on the boat that independently established its longitude and latitude; the signature within the signals that gave the boat's number was in many cases cribbed with help from the Operational Intelligence Centre's tracking files and by putting together a catalogue of each boat's individual radio fingerprint— the background radio noise unique to individual transmitters and the slight individual eccentricities in each radio operator's "fist," the way he varied the length of his Morse code dots and dashes or paused between letters or words.[43]

In late March the first American escort carriers began appearing in the Atlantic; these were small ships that carried twenty-four fighters and navy attack planes and extended air cover across the Atlantic to the convoys. Shipboard direction-finding equipment was now paying off as well; these sets allowed escort destroyers to run down the line of a radio bearing to intercept a U-boat before it got in range of a convoy, increasing by probably 50 percent the number of U-boats sunk by destroyers in February, March, and April according to a study by Rodger Winn's Operational Intelligence Centre. It was air attacks, however, that now were really beginning to tip the balance. In April and May, fifty-six U-boats were destroyed, thirty-three of those in whole or in part the result of air attacks.[44]

On May 19, Dönitz, who earlier in the year had been promoted by Hitler to commander-in-chief of the German navy with the rank of Grossadmiral, broadcast to his commanders an extraordinary do-or-die order:

> If there is anyone who thinks that fighting convoys is no longer possible, he is a weakling and no real U-boat commander. The Battle of the Atlantic gets harder but it is the decisive campaign of the war. Be aware of your high responsibility and be clear that you must answer for your actions. . . . Be hard, draw ahead, and attack. I believe in you. C-in-C.[45]

It took just four more days for reality to sink in even to Dönitz. On May 23 he withdrew his U-boats from the Atlantic convoys and ordered a "temporary shifting of operations to areas less endangered by aircraft," the South Atlantic and the coasts along Gibraltar, Brazil, and Africa. He noted in his diary that not long before he was sinking 100,000 tons of enemy shipping per U-boat lost; that exchange rate was now down to 10,000 tons per U-boat. "Thus losses in May have reached an intolerable level," he acknowledged.[46]

In June, the U.S. Navy's code breakers in Op-20-G did what Dönitz's security investigation earlier in the year had failed to, and tracked to the source their growing suspicion that there had to be a leak somewhere in their own code systems. As early as August 1942, Bletchley Park had warned that German radio messages contained information obtained from reading the Allied convoy code. But GC&CS had only one cryptanalyst assigned to checking the security of their own ciphers and the British naval officials in charge of code production were slow to react and convinced themselves that some stopgap changes introduced in December had mostly fixed the problem.[47] The American code breakers now carefully matched up decrypted Enigma signals advising the U-boats of "expected convoys" with British and American messages that had been transmitted shortly before in Naval Cypher No. 3. It had required a huge fight to get the security-conscious Admiral King to agree to release copies of the American convoy signals to his own navy's investigators, but once they finally got hold of them the match was perfect. The Germans' precise information about convoy movements clearly had come directly from the U.S. Navy's own signals. On June 10 a new convoy code, Cypher No. 5, was ordered into service. For the rest of the war, the Allied convoy signals remained unbroken.[48]

The ASWORG scientists also had been denied access to the Enigma secret on King's orders, but they figured it out themselves—incidentally providing an example of what Dönitz could readily have done had he had an organized scientific staff of his own, which he did not (at least until late in the war, when it was too late to make a critical difference). Morse had assigned Jacinto Steinhardt, a Ph.D. chemist who joined ASWORG in November 1942, to carry out a study of how well Allied radio direction fixes contributed to locating and attacking U-boats. Steinhardt decided to first see how good the data was, and so compared the direction finding reports to the actual locations where U-boats were subsequently located. The DF fixes were unbelievably accurate; in fact, Steinhardt calculated, ten times more accurate than ought to be possible through direction finding alone. Morse and Steinhardt had little trouble deducing the true explanation— that decoded German messages were the real source, which for security reasons were being disguised by the U.S. Navy as DF reports.

At his next meeting with Admiral Low, Morse reported this curious result "with a straight face." He said:

Admiral Low, also with a straight face, said that was interesting. But the next day he called Steinhardt and me in and disclosed what by that time we had guessed but never mentioned, that our side had broken the German code and that the locations given to us as [DF] readings were in fact the positions reported by the submarine skipper himself to his commander.[49]

Enigma decrypts throughout the spring were revealing something of potentially even greater significance than the U-boats' whereabouts: the growing demoralization and even panic of their crews. In his May withdrawal order, Dönitz vowed the battle would soon be rejoined in the North Atlantic, "the enemy's most sensitive area." The U-boat arm, he swore, would soon "subdue the enemy by a continual bloodletting which must cause even the strongest body to bleed to death." But such heroic language was apparently lost on the men under him, now being killed by the thousands a month: Rodger Winn noted in an intelligence report that in April and May, for the first time in the war, U-boat commanders had failed to press home attacks on convoys, even when they were in a favorable position to do so. Enigma decrypts showed that a sharply increasing number of boats were declaring "mechanical difficulties" during their outbound transit of the Bay of Biscay and turning back to port.

Dönitz made the rounds of the bases giving pep talks, but the mood was morose. "No more parties were given to celebrate the start of a campaign now," recalled one commander. "We just drank a glass of champagne in silence and shook hands, trying not to look each other in the eyes."[50]

The battle of the convoys was, in fact, all but over.

IN PRESSING FOR stepped-up air operations in the bay back in March 1943, Blackett had anticipated this development, foreseeing that the withdrawal of the U-boats from the North Atlantic would offer an even more compelling reason to focus the anti-U-boat offensive on the bay; it would also offer an opportunity to deliver the coup de grâce to Dönitz's force. As long as the U-boats remained engaged in attacking the convoys, there was some truth to King's insistence that the "convoy area" was the proper focal point for the application of air power: the convoys were the bait that drew the U-boats into the killing zones of escorting aircraft. But, as Blackett argued with remarkable prescience in a March 23 note to the naval staff, that situation would surely not last. At present the use of VLR aircraft to provide air cover to shadowed convoys was undeniably the best way to employ them; measured in terms of both ships saved and U-boats sunk, each flying hour produced ten times greater returns in that mission than when employed patrolling the bay. But the comparison was "misleading," Blackett cautioned, and certain to change:

> . . . at any time, BDU may find their present offensive against the trans-Atlantic convoys unduly expensive and decide to switch their effort to other areas. Though this would be an important gain for us, we may still lose very heavily by attacks on other convoy routes, independent and in focal areas. It is clearly impossible for us to be equally strong in potential air cover all over the Atlantic (and Indian Ocean) so that "soft" spots must necessarily exist. We would then be forced to switch also our V.L.R. air effort to the threatened area to supplement the normal air cover available there. This is a difficult operation, requiring a large preparatory ground effort, and leaves the initiative in the hands of the enemy. Thus the giving of air cover to shadowed convoys in the N. Atlantic, while extremely profitable now, is unlikely to become a decisive method of defeating the U-boats, since BDU can disengage at will. The advantage of the Bay offensive is that it acts against all U-boats using the Biscay ports, i.e., the great majority of all operational U-boats. BDU cannot disengage from the Bay offensive at will.[51]

Blackett attached an analysis that he and Williams had carried out on the number of VLR aircraft needed to thoroughly cover the bay. It was definitely very much fewer than the numbers that would be needed to protect ships all across the Atlantic and Indian oceans once the U-boats shifted their hunting grounds: "certainly less than 300 aircraft, and probably less than 200." Just forcing the U-boats to stay submerged as much as possible while transiting the bay cut down their operational time at sea significantly. And Blackett noted that even if the U-boats adopted maximum submergence tactics while crossing the bay, they would still have to cross half the distance on the surface, where they were vulnerable to detection by radar and to attack: that was because their submerged speed was so much slower. There was thus a limit to how much the U-boats could ever successfully evade aircraft. Saturating a band about 120 miles wide with constant patrols day and night for twenty-four hours would in theory catch every U-boat passing through that zone during the period.

Even when sightings did not offer a chance to carry out an attack, forcing the U-boats to dive when air patrols passed over produced *increasing* returns on each additional investment in flying time. Each time a boat dived it had to spend an extra quarter of an hour on the surface recharging its batteries just to replace the energy expended diving and resurfacing; that extra time on the surface in turn increased its chances of being spotted by another air patrol. More flying hours thus increased the number of U-boats sighted per flying hour. In March and April, when Coastal Command aircraft began to be equipped in substantial numbers with 10-centimeter radar, Allied aircraft were already spotting 60 percent of the U-boats that made the transit across the bay—which worked out to one for every 30 flying hours—and attacked 40 percent of them.[52]

Those numbers shot up in June and July. Blackett and Williams and others in the Admiralty ORS produced a series of studies on the most effective balance of day and night flying, the advantages of regular versus irregular patrols, the best area and proper width of the patrol patterns, the expected German countermeasures and how best to deal with them. From March to August, aircraft sighted 350 U-boats in the bay and attacked 200 of them. In June, the U-boats began transiting the bay in groups hoping to fight off air attacks with massed antiaircraft fire from their deck guns; the Allies countered by sending surface vessels into the bay to add an additional threat the U-boats had to counter.[53] At the peak of the bay air offensive, every operational U-boat was being sighted on average nearly twice each month by Allied aircraft. From July 2 to August 3, the Allies destroyed 41 more U-boats, 16 of those by air attacks in the bay.[54] The U-boat war, as Admiral Noble had said, was now indeed an air problem: in June through September naval and shore-based aircraft destroyed four times as many U-boats as surface vessels did. So it would remain throughout the rest of the war, with aircraft consistently accounting for one and a half to two times as many kills as surface vessels.[55]

Churchill would reflect in his memoirs on this turning point of the war in the summer of 1943, when "the decisive battle with the U-boats was now fought and won," and the fear and caution yet remained:

> The Battle of the Atlantic was the dominating factor all through the war. Never for one moment could we forget that everything happening elsewhere, on land, on sea, or in the air, depended ultimately on its outcome, and amid all other cares we viewed its changing fortunes day by day with hope or apprehension. . . . Vigilance could never be relaxed. Dire crisis

might at any moment flash upon the scene with brilliant fortune or glare with mortal tragedy.[56]

But hope was in the ascendant even as vigilance remained. One incredible fact stood out: for the month of June 1943, the number of merchant ships lost to U-boats was for the first time no greater than the number of U-boats sunk.[57]

IN JUNE the first of the American four-rotor bombes were completed and running, finally turning the daily U-boat Enigma solution into an almost routine assembly-line process, which would continue to the end of the war as dozens more of the machines were shipped in the ensuing months to

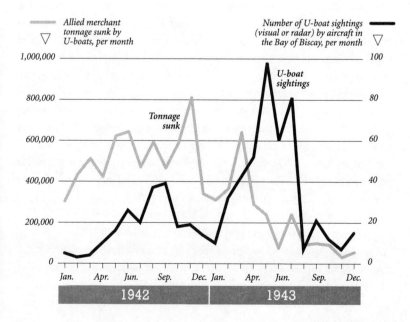

Effects of the Bay offensive

Air attacks on U-boats transiting the Bay of Biscay in the summer of 1943 delivered the decisive blow in the Battle of the Atlantic.

Washington from the National Cash Register factory in Dayton.[58] Bletchley Park would teleprint bombe menus to Washington; the U.S. Navy WAVES who operated the bombes at the Naval Communications Annex, located in a former girls finishing school that the navy had seized at the start of the war in the northwest of the city, set up the machines; and a few hours later the results were on their way back across the Atlantic, and decoded Enigma messages flowed into the Operational Intelligence Centre at the Admiralty and the "Secret Room" within the submarine tracking center at Main Navy in downtown Washington.[59]

To extend the time the U-boats could operate at sea and to minimize their repeated passages through the bay, Dönitz had ordered construction of a dedicated fleet of U-boats that carried large fuel tanks—the "milk cows"—that could refuel the operational boats at sea. Signals specifying rendezvous points were radioed up to two weeks ahead of time. Low argued vigorously for putting this intelligence bonanza to use. Knocking out the tankers, he thought, would be a huge leverage point; each one sunk would pay off in multiples of reduced U-boat days at sea for the entire fleet. There was the chance as well of catching several U-boats in one spot.

The Admiralty sought to veto the plan, fearing that it was too risky: since the actual rendezvous were carried out in radio silence, a successful attack would surely put the Germans on notice that the Enigma had been compromised. Low went ahead anyway. In June, July, and August 1943 the U.S. Navy's escort carriers, now operating in independent task forces, carried out a series of crushing attacks on the tanker rendezvous around the Azores and elsewhere in the Atlantic, sinking 3 of the milk cows and 11 operational U-boats.[60] For the first time in the war the number of operational U-boats at sea fell, dropping by almost half from an average of 120 in May to 60 in August. More important, the effectiveness of each boat plummeted: even when they did operate they accomplished almost nothing. For the rest of the war Dönitz's U-boat fleet barely managed to sink 0.1 merchant vessel per month for each U-boat at sea, down from a peak of forty times that at the height of their prowess. The plots of the Atlantic showing the location of every merchant ship sunk that had been clouds of dark smears covering the Western Approaches of Britain, the east coast of the United States, the ominous "air gap" halfway across the North Atlantic, gave way to a few scattered dots barely noticeable amid a sea of tranquil emptiness: the German submarines, Admiral King announced, had been reduced from a "menace" to a "problem."[61]

"Probably the anti-submarine campaign in 1943 was waged under closer scientific control than any other campaign in the history of the British Armed Forces," Blackett would write later in an appreciation of E. J. Williams.[62] It was, as Churchill had warned, a bitter war "of groping and drowning . . . of ambuscade and stratagem"; but suffusing it all was a war of science. The Allies outfought the Germans at sea: but most of all they outwitted them.

Political Science

ONCE AGAIN a file of U-boats meekly made their way to British ports. Again the sinister silhouettes of an elusive and loathed predator resolved themselves into the contours of a domestic creature, almost pathetic in their helplessness as they chugged along at half speed, wallowing through the waves.

Three weeks before the German surrender on May 8, 1945, the British commander-in-chief, Western Approaches, anticipating the end of five and a half years of gnawing sleepless nights and anxiety-ridden days, sent orders to the fleet specifying procedures to be followed:

On the "Cease Fire" U-Boats will be ordered by the Admiralty:

a. To report position

b. To remain fully surfaced

c. Fly large black or blue flag by day

d. Burn navigation lights by night

e. Jettison ammunition at sea and render all accessible torpedoes safe by removal of pistols. Render mines safe, and remove breech blocks from gun

f. Proceed to harbour or sheltered water for boarding and preliminary inspection

g. To make no signals except in P/L [plain language]

h. To adjust speed to arrive at (f) between sunrise and three hrs. before sunset[1]

Instructions to Allied ships and aircraft directed that they were not to take any offensive actions against a sighted U-boat except if it "is seen to commit a hostile act or shows signs of treachery, for example by diving or failing to comply with the orders."

The cease-fire went into effect at midnight May 8 Central European Time and nine hours later, just after sunup the next morning in the United Kingdom, the first radio signal from a U-boat reporting its location was picked up by Coastal Command. A Sunderland flying boat was immediately diverted to its reported position, about 100 miles northwest of the Irish coast. A little after noon the aircrew sighted a surfaced U-boat, flying a blue and white flag from the conning tower. The crew was crowded on the deck "waving madly and giving the 'thumbs up' sign," the British plane reported. Over the next week 173 more German submarines surrendered. Defiant to the end, 221 German commanders scuttled their boats rather than allow them to fall into Allied hands. Two made it to Argentina.[2]

Dönitz remained unshakable in his loyalty to Hitler to the end. The Kriegsmarine's sole duty, Dönitz declared, was "to stand fanatically behind the National Socialist State . . . each deviation from this is a laxness and a crime." In February 1945, as Germany's military position was collapsing on all fronts, Albert Speer, Hitler's confidant and master planner, drew Dönitz aside after one conference with the Führer and appealed to him desperately that something had to be done to save Germany from complete catastrophe. The Grossadmiral snapped back, "The Führer knows what he is doing."[3]

Hitler rewarded his naval commander's loyalty by naming him his successor. Even in the last days of the Reich, Dönitz issued mad orders to fight on. He issued a secret decree to the naval police, who were even more dreaded than the SS for their summary executions of suspected deserters, to show no mercy to those who failed to do their duty:

We soldiers of the Kriegsmarine know how we have to act. Our military duty, which we fulfill regardless of what may happen to right or left or around us, causes us to stand bold, hard and loyal as a rock of the resistance. A scoundrel who does not behave so must be hung and have a placard fastened to him, "Here hangs a traitor who by his low cowardice allows German women and children to die instead of protecting them like a man."[4]

Succeeding Hitler as the Reich's last Führer on April 30, Dönitz went on the air the next day to exhort the German people to resist not only the Russians on the Eastern Front but the "Anglo-Americans" as well—who, he explained, were now fighting "not for their own peoples but solely for the spreading of Bolshevism in Europe." After the surrender, Dönitz even more improbably imagined the Allies would let him stay in charge of postwar Germany; he wrote Eisenhower on May 15 declaring his intention to hold trials in German courts for those responsible for the concentration camps—which he vehemently insisted no one in the armed forces had known of. Eisenhower did not write back. A week later Dönitz was under arrest. By then he had decided that all the reports about the camps had been "largely exaggerated and were propaganda."[5]

At Nuremberg he and Raeder were tried for war crimes and the prosecution devoted much of its case attempting to prove that Dönitz had issued orders to the U-boat crews to fire on survivors in lifeboats and in the water. The evidence for that was admittedly circumstantial, based on some disputed hearsay and the ambiguities in his written order instructing U-boat commanders to "be severe." The less disputable fact that Dönitz had violated the laws of war by ordering his U-boats to torpedo merchant vessels without warning and not to rescue survivors was effectively countered by the testimony of American officers who acknowledged that U.S. Navy submariners had done exactly the same in the Pacific against Japanese freighters and their crews. Dönitz received a sentence of ten years imprisonment, the lightest punishment imposed by the court.

Dönitz's insistence that his U-boat men had fought "a heroic struggle," from which they emerged "unbroken and unbesmirched," as he declared in a final message radioed to the U-boats just before the surrender, would be echoed by his apologists over the years; they have included more than a few American naval officers who felt their opposing commander had been railroaded at Nuremberg.[6] That sentiment would more disturbingly appear in the latter-day hero worship of U-boat "aces" among a cult of history hobbyists fascinated with everything related to the Third Reich.

Yet even if there were nothing criminal in the German navy's U-boat war—a highly debatable proposition in itself—there was nothing heroic about it at all. It was a squalid and pitiless fight that sent 2,800 ships to the bottom of the cold Atlantic and took the lives of tens of thousands of civilian seamen. On the German side, it was little more than suicide dressed up in Nazi propaganda of sacrifice for the Fatherland. The Third Reich built some 1,150 U-boats; of the 830 that saw operational service during the war, 94 percent were sunk, captured, or lost in fatal accidents by the time of the German surrender. Of the 40,000 men who served on U-boats, 26,000 were killed and 5,000 taken prisoner.[7]

The only other theater that offered similarly appalling odds was the Allied bomber offensive against Germany, which cost the United States and Britain 8,000 large aircraft apiece and took the lives of 76,000 Allied airmen. It also killed an estimated 600,000 German civilians. It was the final irony of the modern industrialized slaughter of the Second World War that the two fronts about which so much romantic and heroic nonsense would be spilled were the most barbaric and pitiless, for the men who fought upon them and their victims alike.[8]

BLACKETT, Zuckerman, Bernal, Watson-Watt, Gordon, and others on the scientific left briefly entertained great expectations that their wartime triumphs had opened the door to the scientifically planned society they had long dreamed of, one in which central planning would organize industry and the economy for the benefit of all. Bernal exultantly proclaimed that the harnessing of science to the war effort had proved everything he had been saying for years: "All that I had thought and written about the possibilities of the ordered utilization of science, I now saw enacted in practice, and I saw that where I had erred was not in overestimating, but in underestimating the constructive power of science," he wrote right after the war.

The key now was to keep that same institutional foothold within government with the return to peace. Bernal thought that science, as it had been employed in the war, would help make socialism palatable by "removing the arbitrary and despotic elements which many persons of genuine liberal feeling imagine to be inherent in all planning." The Association of Scientific Workers enthusiastically issued a report urging the marshaling of production for social needs in peacetime; just as scientific management had supplied the armed services the material they had needed during the war, so "in peace time we shall require the same techniques to study the most efficient ways of utilising the country's resources for the satisfaction of the consumer's needs and desires."[9] At the 1947 annual meeting of the British Association for the Advancement of Science, a session on "Operational

Research in War and Peace" featured Watson-Watt, Zuckerman, Bernal, and other leading scientists who had been a part of the wartime operational research effort calling for wider civilian applications of OR.

It was perhaps inevitable that hopes of duplicating in peacetime the unique wartime success of the scientists who had beat the U-boats would be disappointed. Neither on a personal nor on an institutional level did the wartime camaraderie and enthusiasm survive. Most of those involved in the scientific war against the U-boats quickly drifted away, returning to their old jobs, getting on with lives and careers, forgetting about their brief foray into a world where mathematical equations stood for life and death. A deadpan report by the Admiralty's operational researchers a year after the war's end offered a bit of cynical commentary on their conviction that their contributions had already been forgotten. The report solemnly calculated to several decimal places the proportion of knighthoods, honors, and awards bestowed upon scientists in various branches of government; it found that while 8.6 per 1,000 scientists in the Department of Scientific and Industrial Research had been awarded knighthoods and 22.8 per 1,000 had received lesser honors, the corresponding figures in the Department of Naval Operational Research were 1.0 and 7.0.[10]

With only a handful of exceptions, the British and American scientists literally took up where they left off in their research interests before the war. Six would win Nobel Prizes for their scientific discoveries in physics, chemistry, and physiology. Among them was William Shockley, who had a bizarre and tragic end to a brilliant career: after sharing the Nobel Prize in physics in 1956 for the invention of the transistor, he abysmally failed to make the fortune he might have out of that discovery and grew increasingly withdrawn and obsessive, spending his final years delivering lectures on the racial inheritance of IQ and the intellectual inferiority of blacks.

E. J. Williams, probably the most capable of the operational researchers, survived the war by barely a month, dying from cancer at age forty-two. He had rejoined Blackett at the Admiralty in early 1943 as one of his senior scientists in the Department of Naval Operational Research. Like Blackett he drove himself at a relentless pace; an American visitor found both of the men "tired and exhausted from too many seven day weeks." By April 1945, when he knew he would die, Williams asked his doctor to try to give him as many "effective working days" as possible. "By great will-power over great pain," Blackett recalled, Williams was able to spend his final months that summer completing an article on atomic collision processes for a special issue of *Reviews of Modern Physics* in honor of Niels Bohr's sixtieth birthday. "Throughout E.J.'s illness the most painful element was the sense of wanton frustration and waste—his grief for his work was so acute and it would comfort him to know that others could feel this angle of his tragedy," recalled a close friend. "There was much he wanted to do."[11]

CECIL GORDON WAS one who put his career where his convictions were, becoming the head of a unit the postwar Labour government established in the Board of Trade with the explicit aim of applying operational research to save British industry and expand export markets. His appointment was greeted in the press with headlines declaring HE SHOWED RAF HOW TO DO IT. AND NOW FOR INDUSTRY . . .

At the Board of Trade's Special Research Unit, Gordon launched studies on productivity, consumer needs, the incorporation of operational research into industry. It all went predictably nowhere amid the relieved return to peace and normalcy—and the simple fact that what worked for a centrally organized enterprise where the government was the sole customer had little applicability to the diversity of the consumer marketplace.[12] In 1948 Gordon gave it up, joining the animal genetics faculty at Edinburgh at the invitation of C. H. Waddington, and proceeded to have a miserable falling out with his former Coastal Command ORS colleague. Waddington had set up a sort of left-wing commune in a large country house on the outskirts of the town; ten families shared the house, ate meals together, and carpooled to work. "It provided many anecdotes, usually of a wry kind, and a novel," noted the author of a memorial minute of Waddington for the Royal Society. Gordon was incensed by Waddington's favoritism in the lab and his misuse of university property, and after he gave evidence to an official inquiry into these complaints, he arrived at work a few days later to find he had been moved into a basement room at the genetics institute. The secretary of the university agreed that Gordon was in the right and Waddington in the wrong on almost every matter in the dispute. But he added that Gordon might not spend quite so much of his energy on "smelling out sin."

Gordon then abandoned genetics for community medicine, studying class-based inequalities in health, sickness among Scottish railway workers, and the health care needs of the elderly. Frustrated and unhappy, he died at age fifty-three in 1960. At the time of his death he was chairman of the local Labour Party and representative of the Association of Scientific Workers on the Edinburgh Trades Council, still a committed if now deeply disappointed Marxist.[13]

Blackett also found himself increasingly an odd man out in postwar Britain. His undimmed admiration for the Soviet Union, and a growing antipathy toward American military and foreign policies, had already marked him as a security risk by the summer of 1945. A delegation of thirty British scientists had been invited to Moscow by the Soviet Academy of Science to celebrate the victory over Germany and mark the 220th anniversary of the academy. The night before their scheduled departure on June 14, Blackett, Bernal, and six others were handed back their passports with the exit visas to Russia canceled. An official explanation published in *The Times* the next day stated that the eight "were engaged on work of the greatest importance in the production of war materials and research." Unofficially, there were whispers that Churchill and Cherwell were worried about information on the atomic bomb project being leaked to the Soviets. Blackett, furious, declared he would refuse to participate in any more war work until he received assurances that the government would stop trying to limit his freedom to travel or contact scientists in other countries. "To see Blackett marching out of the Admiralty was a magnificent sight," wrote Nevill Mott, another British physicist (and future Nobel Prize winner) who was barred from making the trip.[14]

Blackett was never as doctrinaire as Bernal, who in 1949 traveled to Moscow, praised Soviet science, and called Lysenko's theories on the inheritance of acquired characteristics a democratic response to "bourgeois" science and a refreshing contrast to the situation in Britain, where "science is in the hands of those who hate peace, whose only aim is to despoil and torture people, so that their own profits can be assured." Bernal airily dismissed reports of the Soviet gulags as "allegations from professed anti-Soviet sources which are unverifiable." On Stalin's death in 1953 Bernal hailed the Soviet dictator, explaining that "the true greatness of Stalin as a leader was his wonderful combination of a deeply scientific approach to all problems with his capacity for feeling and expressing himself in simple and direct human terms."[15] The same year the Soviet Union awarded Bernal the Stalin Peace Prize.

Still, Blackett remained far left and pro-Soviet enough to alarm even the new Labour prime minister, Clement Attlee, who had appointed Blackett to an Advisory Committee on Atomic Energy in 1945 and immediately regretted it. Blackett had succeeded Bernal as president of the Association of Scientific Workers, holding the position from 1943 to 1947, and in an address to the 1945 meeting of the association he called admiring attention to the fact that the Soviet delegation to the International Congress of Trade Unions meeting in London that year included at least three scientists. ("Applause," noted the transcript of his speech at that point.) On the government's atomic energy committee, Blackett wrote a minority report urging a neutral position for Britain and advising against the country's acquiring atomic weapons of its own. That prompted Prime Minister Attlee to respond: "The author, a distinguished scientist, speaks on political and military problems on which he is a layman."[16]

In 1948, the same year Blackett was awarded the Nobel Prize in physics, he published *Military and Political Consequences of Atomic Energy* (the subsequent American edition was titled *Fear, War, and the Bomb*), which strongly endorsed the emerging left-wing credo that America, not the Soviet Union, was now the real threat to world peace. He cast the American-sponsored Baruch Plan to place atomic weapons and atomic energy under international control as a conspiracy to deny the Soviets access to atomic research, and asserted that the atomic bombing of Japan was carried out "not so much as the last military act of the Second World War, but as the first act of the cold diplomatic war with Russia."[17] (Bernal offered an even more rabid view of America and the bomb in an article for the *Daily Worker* several years later in which he made Cherwell the original villain, calling him an "ultra reactionary" whose advocacy of strategic bombing against Germany—"the doctrine of those who hate and despise common people"—now "survives in an infinitely more horrible and militarily futile form in the era of the atom bomb." That and a similar depiction of Cherwell in a 1960 lecture by C. P. Snow on the Tizard-Cherwell rift prompted Zuckerman, who certainly had no love for Cherwell, to counter that he could not "see the Prof as some kind of Second World War Dr. Strangelove.")[18]

Blackett's book earned him a glowing review in *Pravda*, the attention of the FBI, and a place on a list George Orwell supplied the British government naming thirty-eight journalists, writers, and actors who were "crypto-communists, fellow travellers or inclined that way."[19] When Blackett traveled to a scientific conference on cosmic rays in Mexico in 1951 he knew he might have trouble getting a visa to the United States, so arranged to take a Trans-Canadian Air Lines flight from Toronto to Mexico City. On the return trip the plane stopped in Tampa, Florida, to refuel; he and his wife were taken off by U.S. immigration officials, questioned whether they were carrying any subversive literature and what their views on America were, and detained for a few hours before being allowed to depart on the next flight out. Blackett took a slight sardonic amusement from the incident—he subsequently sent his erstwhile jailer a picture postcard from Niagara Falls—but it clearly rankled.[20]

IN BOTH BRITAIN AND AMERICA the war did secure a permanent institutional foothold for scientific advice in government. With the Labour victory in 1945, Henry Tizard returned to Whitehall as the new government's chief scientific adviser. Solly Zuckerman had a distinguished postwar career as the top scientific adviser to the Ministry of Defence in the 1950s and later to the government of Labour prime minister Harold Wilson in the early 1960s. But both were keenly aware of the limitations of scientists as public policy advisers in the business-as-usual world of peace. The great contribution of the operational research scientists to the war effort was, as they often themselves remarked, their ability to ask the right question. But that was within a framework where what constituted success—U-boats sunk, merchant ships saved—was usually clearly understood and widely agreed regardless of preconceived notions or political inclinations. On the great issues of political economy and social policy, though, the question frequently *is* the answer: one can get any scientifically logical and mathematically rigorous answer one wants depending on what criterion of success one chooses to measure.

Blackett often seemed unable to tell the difference. R. V. Jones ran into this right after the war when Blackett chaired a committee to decide on the future of scientific intelligence in the military services. "Blackett had been a hero of my undergraduate days," Jones wrote, and his contributions in the war "had been great." But, Jones continued, "I had seen him make mistakes":

> He tended to jump into a new field, thinking that his fresh ideas were better than those who had worked in the field for some time. Sometimes they were, but not always. He was given to "rational" solutions of problems which sometimes completely overlooked the human aspects involved, and he would then press these solutions with a fervour that belied their apparent rationalism. . . . I always hoped that if the world were collapsing, Blackett and I would find ourselves fighting side by side in the last ditch, but the routes by which we got there would have been very different.

Blackett concluded that each of the three military services as well as MI6, the British secret service, should have its own scientific intelligence staff; even worse, he decided that all of the separate scientific intelligence staffs should be housed together in a new location, away from the existing ministries. It was, Jones argued in vain, an "organizational disaster," the absolute worst of both worlds, both fragmenting the work and isolating it from the service staffs. But Blackett refused to allow any further discussion and dismissed Jones's objections. Jones had promised the members of his scientific intelligence group at the Air Staff he would stay on through "the dull days of peace" to keep the nucleus of their organization together should trouble threaten again, but decided there was now no point and sadly resigned. With the help of recommendations from Cherwell and Churchill, he was named to the chair of natural philosophy at the University of Aberdeen.[21]

Blackett returned to the University of Manchester to head its physics department. A colleague recalled that "some of us certainly thought of him as an admiral" with his air of command, and sometimes harsh intolerance of mistakes. A student remembered being "awestruck by his stately procession down the main stairs for lunch. He always walked in the dead centre of the staircase, disdaining the banisters. He held his hands, naval fashion, in his jacket pockets, with thumbs protruding. He had no nickname: he was Professor Blackett."

In 1953 Blackett moved to Imperial College, London, and in the 1960s served as president of the Royal Society for five years, by most accounts having "mellowed considerably" in his later years. He still was as intensely private and serious a man as ever, however. When a staffer from the American Institute of Physics came to interview him for its oral history project, Blackett refused to allow him to turn on his tape recorder or take any notes, made short work of the list of prepared questions the poor man had brought with him by dismissing each as irrelevant or ill conceived, and then impatiently told him he did not have any more time to waste on the matter. "Why should I tell about my personal life?" he demanded.[22]

IN AMERICA, where the scientists were never quite so political as they were in Britain, there were also hopes that the operational research successes of the war would be the seed of a new scientific discipline that would revolutionize both industry and military science. Phil Morse and George Kimball wrote a textbook drawing on many examples from the antisubmarine war to show how the basic methods of operational research might be applied to other problems. A professional society, the Operations Research Society of America, was established in 1952 and grew to 500 members in its first year; MIT launched a program in OR the same year and began accepting students and was soon studying problems such as easing traffic congestion, scheduling shifts for police and fire departments, and regulating releases of water on the Columbia River dams. For a while there was great enthusiasm in the business world for operational research's promise of quantitative solutions to management problems in organizing workflows, maximizing manufacturing efficiency, and eliminating bottlenecks in production and distribution.[23]

But the glow soon began to fade there, too. In some ways operational research simply became a victim of its own success; ideas that had once been innovative and the special purview of scientifically trained consultants were now commonplace and part of what every business manager learned in MBA programs (as did every military officer at the service academies and war colleges).[24] Yet what was unique and valuable about the wartime operational researchers was in any event precisely that they were *not* professionals, and were doing something never done before. They were some of the most brilliant scientific minds of their generation, out not to make a career of advising the military but to win the war against Hitler. They brought a scientific outlook and a fresh eye to problems that had often been dealt with until then only by tradition, prejudice, or gut feeling.

As the official British history of the scientific contribution to the war observed, it was this more than anything that ultimately defeated Hitler, a man "who had a romantic view of war. . . . Hitler and his generals failed to produce any operational research comparable to the British development. If they had, they would probably have won the submarine campaign and the war."[25]

Few men did more to win that campaign, and that war, than Patrick Blackett, E. J. Williams, and Cecil Gordon. Even Air Marshal Slessor, who had made the crack about strategy by slide rule, paid them an unqualified tribute in a foreword to C. H. Waddington's history of OR in Coastal Command:

A few years ago it would never have occurred to me—or I think to any officer of any fighting Service—that what the R.A.F. soon came to call a

"Boffin," a gentleman in grey flannel bags, whose occupation in life had previously been something markedly unmilitary such as Biology or Physiology, would be able to teach us a great deal about our business. Yet so it was. No one who knows the true facts can have any doubt that a great deal of the credit for what is perhaps still not generally recognised as the resounding victory it was, namely the Battle of the Bay and the defeat of the U-boat in 1943, is due to men like Blackett, Williams, Larnder, Baughan, Easterfield and Waddington.[26]

They did it by an abiding faith in rationality, a basic confidence in the enduring power of arithmetic and simple probability, and a determination to vanquish an evil that they took to heart as a personal duty. Their idealism was all of a piece. If their larger political views were at times utopian, those convictions welled from the same source of rationality, scientific thinking, and acute sense of responsibility for the injustices of the world.

As human beings they were prideful, touchy, opinionated, and sometimes mistaken, human failings too widespread to merit much condemnation. They were also selfless, incorruptible, and absolutely determined to let the facts lead where they will and damn the consequences, human virtues so rare as to seem, at times, almost otherworldly to the men burdened by politics and plans and career ambitions, to whom they showed the way to victory.

Notes

Abbreviations

AAF	Army Air Forces
AIP	American Institute of Physics, Emilio Segrè Visual Archives (College Park, Md.)
A/S	Anti-submarine
ASV	Anti–Surface Vessel radar
ASW	Antisubmarine Warfare
ASWORG	Anti-Submarine Warfare Operations Research Group
CSAWG	Cambridge Scientists' Anti-War Group
CSSAW	Committee for the Scientific Survey of Air Warfare
DF	Direction Finding
GC&CS	Government Code and Cypher School
IWM	Imperial War Museum (London, U.K.)
NARA	National Archives and Records Administration (College Park, Md.)
NDRC	National Defense Research Committee
NSA	National Security Agency
ONI	Office of Naval Intelligence
ORS	Operational Research Section
PRO	Public Record Office, The National Archives of the United Kingdom (Kew, U.K.)
VLR	Very Long Range aircraft

1. An Unconventional Weapon

1. King-Hall, *North Sea Diary*, 229–30.

2. "Twenty U Boats Given Up," *The Times*, November 21, 1918; King-Hall, *North Sea Diary*, 231.

3. King-Hall, *North Sea Diary*, 232–35, 237–38, 241–42.

4. Tarrant, *U-Boat Offensive*, 77.

5. "Deutschland in the Thames," *The Times*, October 14, 1919.

6. Sueter, *Evolution of the Submarine*, 36–38.

7. Budiansky, *Perilous Fight*, 249.

8. "The Escaped Fenians in New-York," *New York Times*, August 20, 1876.

9. Whitman, "Holland"; Morris, *Holland*, 29–42, 50.

10. Morris, *Holland*, 37.

11. Ibid., 46–47.

12. Senate, *Submarine Boat Holland*, 5.

13. Sueter, *Evolution of the Submarine*, 294–95, 303.

14. Ibid., 326–28.

15. Senate, *Submarine Boat Holland*, 6–11.

16. Scheer, *High Sea Fleet*, 12–14; Manchester, *Last Lion*, 1:433–35.

17. Van der Vat, *Atlantic Campaign*, 37.

18. Tarrant, *U-Boat Offensive*, 169–70.

19. Ibid., 7.

20. Ibid., 12.

21. Scheer, *High Sea Fleet*, 36.

22. Ibid., 222–23.

23. Scott, ed., *Diplomatic Correspondence*, 45.

24. Bell, *Blockade of Germany*, 423.

25. Churchill, *World Crisis*, 70.

26. Davies, "Selborne Scheme," 19–23.

27. Manchester, *Last Lion*, 1:437–38, 443; Massie, *Dreadnought*, 405, 408.

28. Davies, "Selborne Scheme," 24, 26, 32–33; Hore, "Blackett at Sea," 55.

2. Cruelty and Squalor

1. "Biographical Notes," Blackett Papers, PB 1/10A; Stevenson, *British Society*, 32–34.

2. Nye, *Blackett*, 16.

3. Blackett, "Education of an Agnostic," 296.

4. "Biographical Notes," Blackett Papers, PB 1/10A.

5. Nye, *Blackett*, 17; Lovell, "Patrick Maynard Stuart Blackett," 3; Alastair Graham to P. M. S. Blackett, November 23, 1967, Blackett Papers, PB 1/11.

6. Hore, "Blackett at Sea," 56–64.

7. Budiansky, *Battle of Wits*, 49–50.

8. Blackett, *Studies of War*, 27.

9. "Extracts from Diary Kept from 1914 to 1916," pp. 3–5, Blackett Papers, PB 1/10A.

10. Gilbert, *First World War*, 252.

11. Ibid., 259.

12. Ibid., 257.

13. Shackleton, *South*, 208.

14. Wells, *Autobiography*, 569.

15. Churchill, *Early Life*, 65.

16. Gilbert, *First World War*, 256.

17. Scott, ed., *Diplomatic Correspondence*, 92.

18. Scheer, *High Sea Fleet*, 248–52.

19. Padfield, *Dönitz*, 10–12, 19–20, 23, 27.

20. Ibid., 53–54, 56, 78; Scheer, *High Sea Fleet*, 253–54.

21. Sims, *Victory at Sea*, 3–4, 7–10.

22. Tarrant, *U-Boat Offensive*, 40; Sims, *Victory at Sea*, 106–11.

23. Sims, *Victory at Sea*, 102–3; Tarrant, *U-Boat Offensive*, 51.

24. Van der Vat, *Atlantic Campaign*, 61–63; Lloyd George, *War Memoirs*, 3:93.

25. Dönitz, *Memoirs*, 4.

26. Tarrant, *U-Boat Offensive*, 69.

27. Terraine, *To Win a War*, 180–84.

28. Ibid., 199, 213.

29. Ibid., 236–37.

30. Ibid., 219, 232n67.

31. Shirer, *Rise and Fall*, 31–32.

3. Cambridge

1. Brown, "Blackett at Cambridge," 97–98.

2. "Professor the Lord Blackett, O.M.," typescript of article for Magdalene College Magazine by "I.A.R." [Ivor A. Richards], Blackett Papers, PB 1/3.

3. Blackett, "Boy Blackett," 11–12.

4. H. E. Piggott to Blackett, February 25, 1957, Blackett Papers, PB 1/11.

5. Cathcart, *Fly in the Cathedral*, 20.

6. Rhodes, *Making of the Atomic Bomb*, 36, 49–50.

7. Clark, *Rise of the Boffins*, 8.

8. Cathcart, *Fly in the Cathedral*, 10–11; Oliphant, *Rutherford*, 19.

9. Brown, "Blackett at Cambridge," 109n18.

10. Quoted in ibid., 106.

11. Slatterly, "Postprandial Proceedings. I," 180.

12. Cathcart, *Fly in the Cathedral*, 112–16.

13. Williamson, ed., *Making of Physicists*, 57.

14. Blackett, "Boy Backett," 9.

15. Brown, "Blackett at Cambridge," 101.

16. "Professor the Lord Blackett, O.M.," Blackett Papers, PB 1/3; Zuckerman, *Six Men*, 13.

17. Bullard, "Blackett."

18. Cathcart, *Fly in the Cathedral*, 42–43.

19. Ibid., 21–22, 115–16; Rhodes, *Making of the Atomic Bomb*, 46, 157.

20. Nye, *Blackett*, 173.

21. Brown, "Blackett at Cambridge," 102.

22. Blackett, "Wilson," 270–71.

23. Brown, "Blackett at Cambridge," 102.

24. Lovell, "Patrick Maynard Stuart Blackett," 12; Bullard, "Blackett."

25. Lovell, "Patrick Maynard Stuart Blackett," 6–10; Brown, "Blackett at Cambridge," 102–3.

26. Bullard, "Blackett"; Zuckerman, *Six Men*, 17.

27. Bullard noted that when teased by left-wing friends about accepting a peerage late in life, Blackett riposted that at least he had remained "Mr. Blackett" throughout his working life.

28. Nye, *Blackett*, 27.

29. Lovell, "Patrick Maynard Stuart Blackett," 11.

30. G. P. S. Occhialini, in Hodgkin et al., "Blackett Memorial," 145.

31. Ibid., 16, 25.

32. Lovell, "Patrick Maynard Stuart Blackett," 76.

33. Wohl, *Generation of 1914*, 223.

34. Sassoon, *Siegfried's Journey*, 160.

35. Stevenson, *British Society*, 103–7.

36. Graves and Hodge, *Long Week-End*, 16.

37. August, *Working Class*, 178–80, 184; Stevenson, *British Society*, 267.

38. Roberts, *A Woman's Place*, 95.

39. Quoted in Wohl, *Generation of 1914*, 292–93n36.

40. Stevenson, *British Society*, 332–33, 415; Graves and Hodge, *Long Week-End*, 42.

41. "The Art of the Jazz," *The Times*, January 14, 1919; "Jazz Dancing—a Canon's Denunciation," *The Times*, March 15, 1919.

42. "Better Plays," *The Times*, August 22, 1919; "A New Shylock," *The Times*, October 10, 1919.

43. Wohl, *Generation of 1914*, 105, 115–16.

44. Stevenson, *British Society*, 95.

45. Manchester, *Last Lion*, 1:791–804; Nye, *Blackett*, 26; "Rush to Aid Government," *New York Times*, May 4, 1926; "Both Sides Are Obstinate," *New York Times*, May 7, 1926.

46. Werskey, *Visible College*, 250; Nye, *Blackett*, 31.

47. Werskey, *Visible College*, 215–16.

48. Blackett, "Frustration of Science," 129.

49. Ibid., 137, 144.

50. Nye, *Blackett*, 27–28; Zuckerman, *Six Men*, 18; Cathcart, *Fly in the Cathedral*, 121.

51. Wohl, *Generation of 1914*, 234.

52. Manchester, *Last Lion*, 2:60–63; Corum, *Luftwaffe*, 76, 115–17.

53. Padfield, *Dönitz*, 107–9.

54. Ibid., 96.

55. Ibid., 101, 111; van der Vat, *Atlantic Campaign*, 83–86.

56. Quoted in Werskey, *Visible College*, 216.

57. Hodgkin et al., "Memorial of Blackett," 145.

58. Ibid., 144.

59. Nye, *Blackett*, 152, 175.

60. Eiduson, *Scientists*, 105–6.

61. "Professor the Lord Blackett, O.M.," Blackett Papers, PB 1/3.

62. Blackett and Occhialini, "Tracks of Penetrating Radiation," 699.

63. Brown, "Blackett at Cambridge," 107; "Atomic Discovery Hailed," *New York Times*, February 18, 1933; "Finds Cosmic Rays Have Odd Particle," *New York Times*, February 17, 1933; Nye, *Blackett*, 50–53.

64. Zuckerman, *Six Men*, 37–38; Brown, "Blackett at Cambridge," 108; Bullard, "Blackett."

65. Nye, *Blackett*, 29, 54.

4. Defiance and Defeatism

1. Werskey, *Visible College*, 339–41.

2. Brown, *J. D. Bernal*, 57–58.

3. Churchill, *Second World War*, 1:71, 73, 77, 111; Manchester, *Last Lion*, 2:100.

4. Manchester, *Last Lion*, 2:102–3.

5. Ibid., 2:92, 145.

6. Ibid., 2:95; Air Ministry, "Expansion of the Royal Air Force," 12; Churchill, *Second World War*, 1:169–70.

7. Manchester, *Last Lion*, 2:99.

8. Wohl, *Generation of 1914*, 105.

9. Fussell, *Great War*, 33, 139; Charles Carrington, quoted in Wohl, *Generation of 1914*, 109.

10. Wohl, *Generation of 1914*, 109, 230.

11. Werskey, *Visible College*, 217–18; "Scientific Workers and War," *Nature*, May 16, 1936, 829–30.

12. Quoted in Bialer, *Shadow of the Bomber*, 158.

13. Douhet, *Command of the Air*, 9.

14. Quoted in Werskey, *Visible College*, 227–28.

15. Quoted in Smith, *British Air Strategy*, 46–47; Bialer, *Shadow of the Bomber*, 14.

16. "Science and Air Bombing," *The Times*, August 8, 1934.

17. Zuckerman, *Six Men*, 20–21; Churchill, *Second World War*, 1:79–80.

18. "Aerial Bombing," *The Times*, August 15, 1934.

19. Rowe, *One Story of Radar*, 4–5.

20. Clark, *Rise of the Boffins*, 15–16.

21. Clark, *Tizard*, 111.

22. Ibid., 116.

23. Ibid., 11.

24. Ibid., 113.

25. Hough and Richards, *Battle of Britain*, 50.

26. "Tizard and the Science of War" in Blackett, *Studies of War*, 105–6.

27. F. A. Lindemann to Lord Swinton, September 23, 1936, Tizard Papers, HTT 111; Churchill to Kingsley Wood, June 9, 1928, PRO, AIR 19/25; Jones, *Wizard War*, 16.

28. Zuckerman, *Six Men*, 20–21; Tizard to D. R. Pye, February 7, 1939, Tizard Papers, HTT 99.

29. Clark, *Tizard*, 127, 141.

30. "Anglo-German Naval Discussions," June 7, 1935, C.P.'s Nos. 86(35) to 140(35), pp. 167–71, PRO, CAB 24/255.

31. Churchill, *Second World War*, 1:140; Padfield, *Dönitz*, 149.

32. Shirer, *Rise and Fall*, 281.

33. Van der Vat, *Atlantic Campaign*, 101; Blair, *Hitler's U-Boat War*, 1:45.

34. Padfield, *Dönitz*, 150–57.

35. Quoted in Zimmerman, "Society for the Protection of Science," 35.

36. Rhodes, *Making of the Atomic Bomb*, 191.

37. Zimmerman, "Society for the Protection of Science," 30–36.

38. Woolf, *Journey Not the Arrival*, 139.

39. Bernal, *Social Function of Science*, 186.

40. Werskey, *Visible College*, 192–93; Baker, "Counter-blast to Bernalism," 174.

41. "Resolution Adopted at the Council Meeting of the Association of Scientific Workers Held on 19th November, 1938," Blackett Papers, PB 5/1/3.

42. Clark, *Rise of the Boffins*, 56–57; Nye, *Blackett*, 35.

43. "Third Progress Report, 3 October 1946," Special Interception Experiments at Biggin Hill, Part I, PRO, AIR 16/179.

44. Larnder, "Origin of Operational Research," 467–71; McCloskey, "Beginnings of OR," 146; Clark, *Rise of the Boffins*, 61–63; Air Ministry, *Operational Research in RAF*, 4–7; Crowther and Whiddington, *Science at War*, 93.

45. "Tizard and the Science of War" in Blackett, *War and Peace*, 103–4.

46. Williams, "Origin of Term," 112; McCloskey, "Beginnings of OR," 146.

47. Quoted in Nye, *Blackett*, 8.

48. Ibid., 38.

49. Shirer, *Rise and Fall*, 471–75.

50. Ibid., 387.

5. Remedial Education

1. Roskill, *War at Sea*, 1:34.

2. Schofield, "Defeat of the U-Boats," 120.

3. Hackmann, "Sonar Research," 87, 88.

4. Padfield, *Dönitz*, 175.

5. Shirer, *Rise and Fall*, 571.

6. Blair, *Hitler's U-Boat War*, 1:46–47.

7. Padfield, *Dönitz*, 150.

8. "Reflections of the C.-in-C., Navy, on the Outbreak of War, September 3, 1939," *Fuehrer Conferences*, 37–38.

9. "Memorandum by Admiral Doenitz, F.O. U-Boats, Date 1.9.39: The Building-Up of the U-Boat Arm," ibid., 36–37.

10. Blair, *Hitler's U-Boat War*, 1:37; Tarrant, *U-Boat Offensive*, 169–76.

11. Padfield, *Dönitz*, 179.

12. Ibid., 180.

13. Gilbert, ed., *Churchill War Papers*, 1:159–60.

14. Ibid., 1:18.

15. "Conference of the C.-in-C., Navy, with the Fuehrer on September 7, 1939," *Fuehrer Conferences*, 39–40.

16. Carroll, "First Shot," 406; van der Vat, *Atlantic Campaign*, 28–30; Churchill, *Second World War*, 1:423.

17. Churchill, *Second World War*, 1:426–28.

18. Ibid., 1:432–34.

19. Gilbert, ed., *Churchill War Papers*, 1:157–59; Churchill, *Second World War*, 1:416.

20. Carroll, "First Shot," 406; "The Athenia," *The Times*, September 5, 1939.

21. Padfield, *Dönitz*, 197–98.

22. Higgins, *Defensively-Armed Merchant Ships*, provides a lucid analysis of the legal issues involved.

23. Shirer, *Rise and Fall*, 639–43; Manchester, *Last Lion*, 2:563–64.

24. Padfield, *Dönitz*, 206.

25. Gilbert, ed., *Churchill War Papers*, 1:235.

26. Miller, *War at Sea*, 43.

27. Padfield, *Dönitz*, 206–7.

28. Gilbert, ed., *Churchill War Papers*, 1:1276.

29. Ibid., 1:637; Manchester, *Last Lion*, 2:568–69.

30. Gilbert, ed., *Churchill War Papers*, 1:722.

31. Ibid., 1:227–28; Churchill, *Second World* War, 1:467–68; Schofield, "Defeat of the U-Boats," 121.

32. Churchill, *Second World War*, 1:465–66.

33. Clark, *Tizard*, 129.

34. Gilbert, ed., *Churchill War Papers*, 1:52–53.

35. Ibid., 1:166.

36. Clark, *Rise of the Boffins*, 72–74.

37. Andrew, *Secret Service*, 453.

38. Denniston Papers, DENN 1/4, p. 5.

39. Mahon, "History of Hut Eight," 14.

40. Hinsley and Stripp, eds., *The Codebreakers*, 77–78, 238.

41. Manchester, *Last Lion*, 2:608–10; Woolf, *Journey Not the Arrival*, 9–10.

42. Padfield, *Dönitz*, 208–11.

43. Tarrant, *U-Boat Offensive*, 83; Blair, *Hitler's U-Boat War*, 1:143–44; *Fuehrer Conferences*, 59.

6. Blackett's Circus

1. Zuckerman, *Apes to Warlords*, 108.

2. Krohn, "Zuckerman," 580, 590, 593.

3. Robertson, "Waddington," 578–79, 603–4.

4. Krohn, "Zuckerman," 577.

5. Huxley, "Science in War."

6. *Science in War*, 11.

7. Larnder, "Origin of Operational Research," 471–72; Zimmerman, "Preparations for War," 118–20; Air Ministry, *Operation Research in RAF*, 6–8, 12, 19.

8. Zimmerman, "Preparations for War," 118.

9. Quoted in Fortun and Schweber, "Scientists and World War II," 613.

10. *Science in War*, 34–35.

11. Werskey, *Visible College*, 264n; Nye, *Blackett*, 36.

12. Padfield, *Dönitz*, 214–15; van der Vat, *Atlantic Campaign*, 190–91.

13. McMahan, "German Navy's Special Intelligence," 21.

14. "U-99 Interrogation of Survivors," April, 1941, pp. 5–6, U-boat Archive, www.uboatarchive.net/U-99INT.htm; "U-70 Interrogation of Survivors," April, 1941, p. 5, U-boat Archive, www.uboatarchive.net/U-70INT.htm.

15. Van der Vat, *Atlantic Campaign*, 252–53.

16. Clark, *Tizard*, 244.

17. Ibid., 228.

18. Jones, *Wizard War*, 23–26.

19. Ibid., 3.

20. Ibid., 94.

21. Ibid., 101; Churchill, *Second World War*, 2:384–85.

22. Jones, *Wizard War*, 109.

23. Clark, *Tizard*, 236–40.

24. Ibid., 253–54.

25. Burns, *Roosevelt*, 1:441.

26. Clark, *Tizard*, 260–69.

27. Morse, *In at the Beginnings*, 121.

28. Bush, *Pieces of the Action*, 33.

29. Burns, *Roosevelt*, 2:60–61; Bush, *Pieces of the Action*, 36.

30. Benson, *Communications Intelligence*, 16–17; Bath, *Tracking Axis Enemy*, 35; Budiansky, *Battle of Wits*, 176–79.

31. Clark, *Tizard*, 214–19; Rhodes, *Making of the Atomic Bomb*, 338.

32. Lovell, "Blackett in War and Peace," 228; Rhodes, *Making of the Atomic Bomb*, 368–69. After Tizard's CSSAW was disbanded in June 1940, Thomson's atomic bomb committee was transferred to the Ministry of Aircraft Production, and was known as the Maud Committee.

33. Overy, *Battle of Britain*, 80; Shirer, *Rise and Fall*, 778.

34. Lovell, "Patrick Maynard Stuart Blackett," 56; Lovell, "Blackett in War and Peace," 222.

35. "Recollections of Problems Studied," in Blackett, *Studies of War*, 207–8.

36. Sir Fredrick Alfred Pile, *Oxford Dictionary of National Biography*; Pile, *Ack-ack*, 26, 113, 160.

37. Clark, *Rise of the Boffins*, 143; I. Evans, "The Beginnings of Operational Research," Blackett Papers, PB 4/7/3/20.

38. "Recollections of Problems Studied," in Blackett, *Studies of War*, 208; George Johnston, WW2 People's War archive, www.bbc.co.uk/ww2peopleswar/stories/28 /a8644728.shtml; E. E. Treadwell, in Sawyer et al., "Reminiscences," 133–35.

39. "Recollections of Problems Studied," in Blackett, *Studies of War*, 209–11.

40. "Operational Research. Professor P. M. S. Blackett," pp. 6–7, Blackett Papers, PB 4/7/2/7; "Recollections of Problems Studied," in Blackett, *Studies of War*, 220.

41. Churchill, *Second World War*, 2:560.

42. Clark, *Rise of the Boffins*, 150; "Formation and Organisation of Naval Operational Research Department," PRO, ADM 1/20113; Lovell, "Blackett in War and Peace," 223.

7. The Real War

1. Blackett, "Evan James Williams," 387, 400–401, 403–4.

2. "Coastal Command Against the U-Boat, 1939–1944," p. 1, PRO, AIR 15/291; "Memorandum on Anti-Submarine Measures," September 1941, Table III, Blackett Papers, PB 4/7/1/6.

3. "Captain A/S., Belfast, to Command-in-Chief, Western Approaches, August 18, 1940," PRO, AIR 15/29.

4. Price, *Aircraft Versus Submarine*, 47–48.

5. Churchill, *Second World War*, 3:100–101, 111–12, 666.

6. Ibid., 3:663.

7. Price, *Aircraft Versus Submarine*, 60–61.

8. "Scientists at the Operational Level by P. M. S. Blackett, Notes on Statement to Admiralty Panel on 16.9.41 on the Function of the Operational Research Section (O.R.S.)," Blackett Papers, PB 4/7/1/2.

9. "Coastal Command Tactical Instruction No. 12. Attack on U-Boats with Depth Charges Dropped from Aircraft," January 27, 1941, PRO, AIR 15/29.

10. "Analysis of Attacks on U-Boats by Aircraft," ORS/CC/5/8 Report No. 142, E. J. Williams, September 11, 1941, PRO, AIR 15/731.

11. "Memorandum on Anti-Submarine Measures," September 1941, Blackett Papers, PB 4/7/1/6.

12. Waddington, *OR in World War II*, 172; "Coastal Command Tactical Instruction No. 18 Attacks on Submarines," December 15, 1941, Folder: FX-40, A16(2) Procedures, Attacks, Aircraft—(Doctrines, Tactics, Etc.) 1941–1942, NARA, Tenth Fleet: Subject Files, Box 3. Coastal Command apparently began to implement Williams's recommendations even before his formal report was completed; in typewritten notes Blackett later made about Williams's war work he added by hand, "Action preceded appearance of all these reports." See "Summary of War Work," Blackett Papers, PB 8/10/2.

13. "Recollections of Problems Studied," in Blackett, *Studies of War*, 216.

14. Ibid., 216–17; "Note on the Camouflage of Aircraft Engaged on Anti-Submarine Operations," Report No. O.R.S.C.C./144, by W. R. Merton, PRO, AIR 15/731; "Memorandum on Anti-Submarine Measures," September 1941, Blackett Papers, PB 4/7/1/6.

15. "Notes for Lecture on Science and the U-Boat War, D.N.O.R.—May 1943," pp. 9–10, Blackett Papers, PB 4/7/1/5. (A copy of this same document is also in PRO, ADM 219/629.)

16. Morse and Kimball, *Methods of Operations Research*, 3–4.

17. "Operational Research. Notes of a Lecture Given by Professor P.M.S. Blackett, F.R.S., to Institute of Physics on 17th March 1953, Sydney, Australia," pp. 6–7, Blackett Papers, PB 4/7/2/9.

18. Llewellyn-Jones, "Case for Large Convoys," 139.

19. "Scientists at the Operational Level by P. M. S. Blackett, Notes on Statement to Admiralty Panel on 16.9.41 on the Function of the Operational Research Section (O.R.S.)," Blackett Papers, PB 4/7/1/2.

20. Christopherson and Baughan, "Reminiscences," 574; "Monograph on the Work of the Operational Research Section Coastal Command, edited by C. H. Waddington, Sc. D.," Staff of the Operational Research Section, Coastal Command (front matter), PRO, AIR 15/133; Clark, *Rise of the Boffins*, 144.

21. Llewellyn-Jones, "Case for Large Convoys," 162n50.

22. Sawyer et al., "Reminiscences," 129.

23. Calder, *Science in Our Lives*, 143–44.

24. Middlebrook, *Convoy*, 109.

25. Lawrence, *Tales of the North Atlantic*, 172.

26. Middlebrook, *Convoy*, 105; Brown, *Atlantic Escorts*, 48–50.

27. Monsarrat, *Cruel Sea*, 7, 46, 87, 104.

28. Ibid., 33, 144.

29. Miller, *War at Sea*, 178; Middlebrook, *Convoy*, 110.

30. Buchheim, *Das Boot*, 11.

31. Ibid., 55–56, 282; Blair, *Hitler's U-Boat War*, 1:57–64.

32. "U-70 Interrogation of Survivors," April 1941, pp. 10–11, U-boat Archive, www.uboatarchive.net/U-70INT.htm.

33. Kahn, *Seizing the Enigma*, 154–59.

34. "Report, Capture of U 110, Commander A. Baker-Cresswell," May 10, 1941, PRO, ADM1/11133; Balme, Oral History.

35. Erskine, "Naval Enigma: A Missing Link," 497.

36. Hinsley and Stripp, eds., *Codebreakers*, 134.

37. "Enigma General Procedure," NARA, Historic Cryptographic Collection, NR 1679.

38. Kahn, *Seizing the Enigma*, 176.

39. Milner-Barry, Oral History.

40. Frank Birch to Edward Travis, August 15, 1941, PRO, HW 14/18.

41. Alastair Denniston, August 18, 1941, PRO, HW 14/18.

42. Alfred Dillwyn Knox to Alastair Denniston, November 1941, PRO, HW 14/22.

43. Milner-Barry, "Action This Day."

44. McMahan, "German Reactions," 186.

45. "Conference of the C.-in-C., Navy, with the Fuehrer in the Afternoon of June 21,

1941," *Fuehrer Conferences*, 220; "Conference of the C.-in-C., Navy, with the Fuehrer in Wolfsschanze in the Afternoon of July 25, 1941," ibid., 222.

46. Burns, *Roosevelt*, 2:105.

47. Ibid., 2:140–41; Morison, *History of Naval Operations*, 1:84–85.

48. Gilbert, *Churchill and America*, 239.

49. Klein, *Woody Guthrie*, 142.

50. Ibid., 209–10.

8. Baker's Dozen

1. Churchill, *Second World War*, 3:606–8.

2. "Report of the C.-in-C., Navy, to the Fuehrer in Berlin on December 12, 1941," *Fuehrer Conferences*, 245.

3. Sternhell and Thorndike, *Antisubmarine Warfare*, 84.

4. Padfield, *Dönitz*, 237–38.

5. "Manuscript for Division 6 Long History," p. 3-1, NARA, Division 6 Records, Box 42; "Steps in the Organization of Section C-4 NDRC," ibid., Box 39; NDRC, *Survey of Subsurface Warfare*, 21–22, 26–27.

6. "Material Brought Back from London by Dr. Tate," NARA, Division 6 Records, Box 59.

7. Morse, *In at the Beginnings*, 29, 55, 65, 125, 134–35, 137.

8. Ibid., 118, 137.

9. Ibid., 16, 138, 153–55; Feshbach, "Philip McCord Morse," 250.

10. Morse, *In at the Beginnings*, 170.

11. Telephone interview with Wilder D. Baker, Jr., July 2, 2011.

12. "Resume of Anti-Submarine Operations Against the German U-Boats in World War II," p. 2, Folder: Tenth Fleet, NARA, Tenth Fleet: Subject Files, Box 34.

13. "Formation and Organisation of Naval Operational Research Department," PRO, ADM 1/20113; Lovell, "Blackett in War and Peace," 224.

14. "Memorandum for Rear Admiral J. A. Furer, U.S.N., March 16, 1942; Subject: Formation of Statistical and Analytical Groups Under Anti-Submarine Warfare Unit," ASWORG, *Review of Activity*, 48.

15. "From: Atlantic Fleet Anti-Submarine Warfare Officer, To: Coordinator of Research and Development (Secretary's Office); Subject: Records and Analyses of Anti-submarine Warfare, March 16, 1942," ASWORG, *Review of Activity*, 45–47.

16. Morse, *In at the Beginnings*, 174.

17. "Organization Report of NDRC Operational Research Group with the U.S. Navy A.S.W. Committee, April 2, 1942," Folder: ASW General Organization, NARA, Tenth Fleet: Subject Files, Box 24; "ASWORG Personnel," ASWORG, *Review of Activity*, 59–60; Horvath and Ernst, "Morse: A Remembrance," 8; Hickman and Heacox, "Actuaries."

18. Love, "Ernest Joseph King," 140.

19. Buell, *Master of Sea Power*, 123.

20. Ibid., 146.

21. Ibid., 124, 222–23.

22. Ibid., 284, 521; Cohen and Gooch, *Military Misfortunes*, 85.

23. Cohen and Gooch, *Military Misfortunes*, 78–79; Craven and Cate, eds., *Army Air Forces*, 1:528–30.

24. Layton et al., "*And I Was There*," 140.

25. PRO, ADM 223/286.

26. "A Study of the U-Boat Campaign, Atlantic Fleet ASW Unit," p. 6, Folder 6: The U/B Campaign, NARA, Tenth Fleet: ASW Series II; "Atlantic Theater 1942, Op-16-C, January 12, 1943," ibid.; Meigs, *Slide Rules and Submarines*, 53.

27. Miller, *War at Sea*, 306.

28. Buell, *Master of Sea Power*, 289–90.

29. "From: The Secretary, Operational Proposals Board, To: Captain Wilder D. Baker, Subject: Anti-submarine Warfare—Suggestion and Criticisms from the Public Regarding, General Classes Of, June 1, 1942," Folder: ASW General Organization, NARA, Tenth Fleet: Subject Files, Box 24.

30. Van der Vat, *Atlantic Campaign*, 347.

31. Tarrant, *U-Boat Offensive*, 106–7.

32. Cohen and Gooch, *Military Misfortunes*, 87.

33. Padfield, *Dönitz*, 241.

34. "Resume of Anti-Submarine Operations Against the German U-Boats in World War II," p. 9, Folder: Tenth Fleet, NARA, Tenth Fleet: Subject Files, Box 34; Meigs, *Slide Rules and Submarines*, 53.

35. "Memorandum of Conference Between Representatives of the Atlantic Fleet ASW Unit and Representatives of the Commander-in-Chief, United States Fleet, ASW Unit, April 28, 1942," Folder: ASW General Organization, NARA, Tenth Fleet: Subject Files, Box 24.

36. "Memorandum for: Admiral King, June 24, 1942, The Battle of the Atlantic Is Being Lost," Folder: Anti-Submarine Measures (Appreciation & Summary), NARA, Tenth Fleet: Subject Files, Box 27.

37. Morse, *In at the Beginnings*, 176–77.

38. "Preliminary Report on the Submarine Search Problem, May 1, 1942," Folder: Submarines B, NARA, Records of Edward L. Bowles.

39. Morse, *In at the Beginnings*, 179–80.

40. "Comments on the Organization of Operations Research by Philip M. Morse," pp. 1–2, Folder: A10(12). FX-40 Miscellaneous Publications, 1942, NARA, Tenth Fleet: Subject Files, Box 2.

41. "Instructions for ASWORG Members, December 1, 1942," p. 1, Folder 27: N.D.R.C. & A.S.W.O.R.G. (Organization), NARA, Tenth Fleet: ASW Series II.

42. "From: Lieutenant Commander A. B. Vosseller, To: The Commanding Officer, Atlantic Fleet Anti-Submarine Warfare Unit, Subject: Report on Inspection Trip, April 13, 1942," Folder: ASW General Organization, NARA, Tenth Fleet: Subject Files, Box 24; "Base Assignments," Appendix E, ASWORG, *Review of Activity*, 61; Morse, *In at the Beginnings*, 180–81.

43. ASWORG, *Review of Activity*, 10.

44. Smyth, "Oral History of Dr. Ralph Beatty, Jr.," 57; "Progress Report Covering Period September 15–October 15, 1941," p. 1, Folder 27: N.D.R.C. & A.S.W.O.R.G. (Organization), NARA, Tenth Fleet: ASW Series II.

45. Brinkley, *Washington Goes to War*, 106, 241–43.

46. ASWORG, *Review of Activity*, 22–27.

47. Morse, *In at the Beginnings*, 182. "Memoranda (Reports) Prepared by ASW Operations Research Group, Sec. 6.1, NDRC Coordinating with Anti-Submarine Forces—U.S. Army and Navy," Folder 27: N.D.R.C. & A.S.W.O.R.G. (Organization), NARA, Tenth Fleet: ASW Series II, has a list of all ASWORG reports and the abstract for Report No. 11 notes: "A quantitative treatment of the problem along these lines was reported in the British ORS/CC/5/S Report #142." Exactly when the ASWORG scientists became aware of the British report is not clear.

48. "Progress Report Covering Period September 15–October 15, 1941," p. 1, Folder 27: N.D.R.C. & A.S.W.O.R.G. (Organization), NARA, Tenth Fleet: ASW Series II.

49. "Tentative Doctrine for Anti-Submarine Warfare by Aircraft, October 17, 1942," Folder: A16(2) FX-40, Procedures, Attacks, Aircraft (Doctrines, Tactics, etc.), 1941–42, NARA, Tenth Fleet: Subject Files, Box 3.

9. Closing the Gaps

1. Morison, *History of Naval Operations*, 1:241; Craven and Cate, eds., *Army Air Forces*, 1:537.

2. Bennett, *Behind the Battle*, 154; Mark, *Aerial Interdiction*, 230.

3. Ray, *Night Blitz*, 105.

4. Quoted in Jones, *Wizard War*, 183.

5. Bennett, *Behind the Battle*, 148; Smith, *British Air Strategy*, 64.

6. Kennett, *History of Strategic Bombing*, 133.

7. Bennett, *Behind the Battle*, 148, 170; Buckley, *Air Power*, 164.

8. Hinsley et al., *British Intelligence*, 2:636.

9. Craven and Cate, eds., *Army Air Forces*, 1:539; van der Vat, *Atlantic Campaign*, 385, 387.

10. "Mid-May Situation of A/S AC Squadrons," Folder 8: ASM Air Forces and Strengths, NARA, Tenth Fleet: ASW Series II; "Memorandum on Anti-Submarine Measures," September 1941, Table III, Blackett Papers, PB 4/7/1/6.

11. Van der Vat, *Atlantic Campaign*, 386–87.

12. Webster and Frankland, *Strategic Air Offensive*, 4:205–13; Clark, *Tizard*, 308.

13. "Quantitative Study of Total Effects of Air Raids (Hull and Birmingham Survey), Ministry of Home Security Research and Experiments Department, April 8, 1942," reprinted in Zuckerman, *Apes to Warlords*, Appendix 2; Clark, *Tizard*, 309; Crook, "Case Against Area Bombing," 181; "Estimation of Bombing Effect," H. T. Tizard, p. 2, Blackett Papers, PB 4/4/8.

14. Crowther and Whiddington, *Science at War*, 98; "Effect of Bombing Policy," Blackett Papers, PB 4/4/7; "Recollections of Problems Studied," in Blackett, *Studies of War*, 224.

15. Clark, *Tizard*, 309.

16. Ibid., 310, 313.

17. Webster and Frankland, *Strategic Air Offensive*, 1:332.

18. "Science and Government," in Blackett, *Studies of War*, 124.

19. J. H. Godfrey to 1st Sea Lord, April 8, 1942, Blackett Papers, PB 4/4/6; "Effect of Bombing Policy," Blackett Papers, PB 4/4/7.

20. "Recollections of Problems Studied," in Blackett, *Studies of War*, 226–27.

21. Crook, "Case Against Area Bombing," 174.

22. "Tizard and the Science of War," in Blackett, *Studies of War*, 110.

23. Blackett Papers, PB 4/7/1/14.

24. "The Government and the War," Blackett Papers, PB 4/6/1.

25. "Science and Government," in Blackett, *Studies of War*, 123, 125–26.

26. "The Strategic Air War Against Germany, British Bombing Survey Unit," p. 27, PRO, AIR 10/3866.

27. "Summary of War Work," Blackett Papers, PB 8/10/2; Rosenhead, "Operational Research at the Crossroads," 8–9; "Monograph on the Work of the Operational Research Section Coastal Command, Edited by C. H. Waddington, Sc. D.," Staff of the Operational Research Section, Coastal Command (front matter), PRO, AIR 15/133.

28. Rosenhead, "Operational Research at the Crossroads," 4–5, 23; T. E. Easterfield, in Sawyer et al., "Reminiscences of OR," 127.

29. Waddington, *OR in World War II*, 77.

30. C. Portal, chief of Air Staff, to Prime Minister, July 5, 1942, PRO, ADM 1/12125; Prime Minister's Personal Minute to First Lord and Secretary of State for Air, July 14, 1942, ibid.

31. Waddington, *OR in World War II*, 53–55, 62–63; Rosenhead, "Operational Research at the Crossroads," 9; "Operational Research, Professor P. M. S. Blackett," p. 3, Blackett Papers, PB 4/7/2/7.

32. T. E. Easterfield, in Sawyer et al., "Reminiscences of OR," 127, 129.

33. "Measures to Improve A/S Air Offensive," ORC/CC/Report 174, E. J. Williams, February 12, 1942, PRO, AIR 15/731.

34. "Notes for Lecture on Science and the U-Boat War, D.N.O.R.—May 1943," pp. 10–11, Blackett Papers, PB 4/7/1/5; Price, *Aircraft Versus Submarine*, 51–54.

35. Price, *Aircraft Versus Submarine*, 54–60.

36. "Measures to Improve A/S Air Offensive," ORC/CC/Report 174, February 12, 1942, E. J. Williams, PRO, AIR 15/731.

37. Waddington, *OR in World War II*, 231.

38. Ibid., 36.

39. Tarrant, *U-Boat Offensive*, 105, 106, 109.

40. Van der Vat, *Atlantic Campaign*, 407.

41. Padfield, *Dönitz*, 252–55.

42. Ibid., 256, 326.

43. Ibid., 248.

44. Price, *Aircraft Versus Submarine*, 88–89.

45. "Air Offensive Against U-boats in Transit," October 11, 1942, E. J. Williams, PRO, AIR 15/732; "Notes for Lecture on Science and the U-Boat War, D.N.O.R.—May 1943," pp. 11–12, Blackett Papers, PB 4/7/1/5; Waddington, *OR in World War II*, 232–33.

46. McMahan, "German Navy's Special Intelligence," 68–69.

47. Whitehead, "Cobra and Other Bombes."

48. "Captain Wenger Memoranda," NARA, Historic Cryptographic Collection, NR 4419; Erskine, "Holden Agreement"; Budiansky, *Battle of Wits*, 237–39.

49. Hinsley et al., *British Intelligence*, 2:548.

50. Mahon, "History of Hut Eight," 62; Hinsley et al., *British Intelligence*, 2:747.

51. Mahon, "History of Hut Eight," 74.

52. Kahn, *Seizing the Enigma*, 223–26.

53. Erskine, "Naval Enigma: Heimisch and Triton," 170; Goodman, McMahan, and Carpenter, "Battle of the Atlantic," 269–70; Erskine, "Kriegsmarine Short Signal Systems," 82.

54. "Notes for Lecture on Science and the U-Boat War, D.N.O.R.—May 1943," pp. 13–14, Blackett Papers, PB 4/7/1/5.

55. McMahan, "German Navy's Special Intelligence," 103–4; McMahan, "German Reactions," 199–207.

56. Padfield, *Dönitz*, 274.

57. Schoenfeld, *Stalking the U-boat*, 20–25.

58. Ibid., 25–26; Budiansky, *Air Power*, 309.

59. "To: Dr. E. L. Bowles, From: Dr. P. M. Morse, Subject: General Impressions Gained During a Visit to England, January 4, 194[3]," p. 4, Box 31, Folder 7, Bowles Papers. This memorandum is misdated 1942.

60. "Diary of Dr. Philip M. Morse and Dr. William Shockley, Visit to London Commencing November 19, 1942," pp. 1, 3, 5–9, 15, Box 39, Folder 5, Bowles Papers; Morse, *In at the Beginnings*, 191–92.

61. "General Impressions," p. 8, Box 31, Folder 7, Bowles Papers.

62. "Memorandum to: Brig. Gen. H. M. McClelland, Director of Technical Services, Headquarters, Army Air Forces, Subject: Civilian Specialists in Combat Zones, September 3, 1942," Box 29, Folder 6, Bowles Papers.

63. "Diary of Dr. Philip M. Morse," pp. 17, 23, 28, 32, 38, Box 39, Folder 5, Bowles Papers; Morse, *In at the Beginnings*, 197; "Base Assignments," Appendix E, ASWORG, *Review of Activity*, 62.

64. Schoenfeld, *Stalking the U-boat*, 32.

65. Morse, *In at the Beginnings*, 196.

10. A Very Scientific Victory

1. Llewellyn-Jones, "Case for Large Convoys," 140, 142; Clark, *Rise of the Boffins*, 162.

2. "Recollections of Problems Studied," in Blackett, *Studies of War*, 228.

3. "Statistical Analysis of Effect of Surface Escort of Convoys," Report No. 21/43, p. 6, PRO, ADM 219/37 (also in Blackett Papers, PB 4/5/2); "Note on Variation of Losses with Size of Convoy," Report No. 30/43, PRO, ADM 219/46.

4. "Recollections of Problems Studied," in Blackett, *Studies of War*, 231.

5. "Progress of Analysis of the Value of Escort Vessels and Aircraft in the Anti-U-boat Campaign, Report by Professor Blackett, February 5, 1943," Blackett Papers, PB 4/5/1.

6. "The Case for Large Convoys," Report No. 2/43, Appendix I, PRO, ADM 219/19.

7. Ibid., pp. 1–2; Llewellyn-Jones, "Case for Large Convoys," 149.

8. "Tizard and the Science of War," in Blackett, *Studies of War*, 115.

9. "The Case for an Immediate, Even If Temporary Increase in the Size of Convoy, C.A.O.R. 5.3.43," Blackett Papers, PB 4/5/1.

10. "Recollections of Problems Studied," in Blackett, *Studies of* War, 231–34; Llewellyn-Jones, "Case for Large Convoys," 144.

11. Craven and Cate, eds., *Army Air Forces*, 6:498, 562.

12. Fussell, *Wartime*, 132.

13. Stiles, *Serenade*, 141–42.

14. Konvitz, "Bombs, Cities, Submarines," 29–30.

15. Craven and Cate, eds., *Army Air Forces*, 2:296–97; Konvitz, "Bombs, Cities, Submarines," 40.

16. Konvitz, "Bombs, Cities, Submarines," 27–28.

17. Ibid., 28–29.

18. "Perforability of German Submarine Pens, Located at Lorient and Elsewhere, by American and British Aerial Bombs, ORS Special Report No. 1, Headquarters VIII Bomber Command, December 8, 1942," NARA, Records of Edward L. Bowles.

19. Konvitz, "Bombs, Cities, Submarines," 25, 31.

20. Craven and Cate, eds., *Army Air Forces*, 2:248–49.

21. Eaker, Oral History, p. 6.

22. Konvitz, "Bombs, Cities, Submarines," 33; Craven and Cate, eds., *Army Air Forces*, 2:305–6.

23. Konvitz, "Bombs, Cities, Submarines," 34–36, 43; Craven and Cate, eds., *Army Air Forces*, 2:315–16.

24. "A.U.(43) 96, 29th March, 1943, The Battle of the Atlantic, Memorandum by the Air Officer Commanding-in-Chief, Bomber Command," and "Appendix A, Memorandum by A.O.C.-in-C., Bomber Command," Blackett Papers, PB 4/4/11.

25. Price, *Aircraft Versus Submarine*, 109–10, 161.

26. Bennett, *Behind the Battle*, 154.

27. "Bombing of U-Boat Bases, Note by C.A.O.R.," Prof. Blackett, [March 30, 1943?], Blackett Papers, PB 4/4/11.

28. "Appendix A, Memorandum by A.O.C.-in-C., Bomber Command," Blackett Papers, PB 4/4/11.

29. Slessor, *Central Blue*, 524–25.

30. Llewellyn-Jones, "Case for Large Convoys," 152.

31. "Atlantic Convoy Conference, Washington D.C., March 1 to March 12, 1943, Report of Conference," p. 4, NARA, Records of Edward L. Bowles.

32. "Memorandum to: General Marshall, Lieut. General McNarney. Subject: Recommendations—Army Air Antisubmarine Effort, March 3, 1943," pp. 1–2, Folder: Tenth Fleet—Organization, NARA, Tenth Fleet: Subject Files, Box 34; Meigs, *Slide Rules and Submarines*, 92.

33. "A.C.C.—3/1, Atlantic Convoy Conference, Air Support, Recommendation and Comment by Sub-Committee on Coordination and Implementation, March 12, 1943," NARA, Records of Edward L. Bowles.

34. "Memorandum to: The Secretary of War, Subject: Antisubmarine Aircraft, April 2, 1943," Box 30, Folder 1, Bowles Papers.

35. Craven and Cate, eds., *Army Air Forces*, 2:387.

36. Schoenfeld, *Stalking the U-boat*, 172. An earlier sinking credited to the group was the result of a mistaken identification: on February 10 one of the Liberators attacked a U-boat believed to be *U-519*, which was shortly thereafter reported lost in the Bay of Biscay. In fact the boat that the aircraft attacked (and only slightly damaged) was *U-752*; the cause of *U-519*'s disappearance remains unknown, and was possibly accidental.

37. "Air Offensive Against the U-Boat, Memorandum by the Chief of Staff, U.S. Army, April 19, 1943," Folder: Tenth Fleet—Organization, NARA, Tenth Fleet: Subject Files, Box 34.

38. "Memorandum for the Secretary, April 6, 1943," Box 32, Folder 1, Bowles Papers; Arthur L. Kip to Edward L. Bowles, June 4, 1943, ibid.

39. Buell, *Master of Sea Power,* 242, 252.

40. Craven and Cate, eds., *Army Air Forces,* 2:408–9.

41. Morse, *In at the Beginnings,* 184; Meigs, *Slide Rules and Submarines,* 95.

42. ASWORG, *Review of Activity,* 29–31, 59–60; Morse, "George Elbert Kimball," 136–37.

43. Erskine, "Kriegsmarine Short Signal Systems," 84.

44. "On the Value of Squash in Pack Attacks," Operational Intelligence Centre Special Intelligence Summary O.I.C./S.I. 1254, PRO, ADM 223/261; Tarrant, *U-Boat Offensive,* 118–19.

45. Padfield, *Dönitz,* 299.

46. Ibid., 299–300.

47. Erskine, "Admiralty and Cipher Machines," 49.

48. NSA, *Battle of the Atlantic,* 3:44–56.

49. Morse, *In at the Beginnings,* 185–87.

50. Padfield, *Dönitz,* 301, 318; Hinsley et al., *British Intelligence,* 2:567–68.

51. "Note on the Relation Between the Use of Aircraft to Give Air Cover to Convoys and in the Bay, P. M. S. Blackett, C.A.O.R., 22nd March 1943," pp. 213–14, PRO, ADM 205/30.

52. "Notes for Lecture on Science and the U-Boat War, D.N.O.R.—May 1943," pp. 14–15, Blackett Papers, PB 4/7/1/5.

53. "Brief History of the Bay Offensive, and Discussion of the Present Situation," Report No. 45/43, E. J. Williams, October 9, 1943, PRO, ADM 219/61; "Operational Research in War," Report No. 1/45, Leon Solomon, October 29, 1945, pp. 3–4, PRO, ADM 219/208.

54. Lovell, "Blackett in War and Peace," 226; Craven and Cate, eds., *Army Air Forces,* 2:396.

55. Tarrant, *U-Boat Offensive,* 124, 131, 142.

56. Churchill, *Second World War,* 5:6–7.

57. Tarrant, *U-Boat Offensive,* 120, 123.

58. Budiansky, *Battle of Wits,* 295.

59. Parsons, *Functions of "Secret Room."*

60. Erskine, "Ultra and U.S. Carrier Operations"; Hinsley et al., *British Intelligence,* 2:549, 3(1):213–14.

61. Tarrant, *U-Boat Offensive,* 122; Sternhell and Thorndike, *Antisubmarine Warfare,* 84; "Navy Is Confident," *New York Times,* April 24, 1944.

62. Blackett, "Evan James Williams," 399.

11. Political Science

1. "Surrender of the U-Boat Fleet at Sea. (Short Title—Pledge One.)," April 19, 1943, PRO, AIR 15/449.

2. "From: H.Q.C.C., To: All Coastal Command Groups," May 7, 1945, PRO, AIR 15/449; "Surrender of U/Boats 1945," ibid.; Blair, *Hitler's U-Boat War*, 2:815–19; van der Vat, *Atlantic Campaign*, 531.

3. Miller, *War at Sea*, 502; Padfield, *Dönitz*, 385.

4. Padfield, *Dönitz*, 392, 415.

5. Ibid., 428–34; Shirer, *Rise and Fall*, 1138.

6. Van der Vat, *Atlantic Campaign*, 530; Blair, *Hitler's U-Boat War*, 2:705.

7. Tarrant, *U-Boat Offensive*, 151; van der Vat, *Atlantic Campaign*, 531.

8. Werrell, "Strategic Bombing," 708.

9. Werskey, *Visible College*, 273–74.

10. "Note on the Scale of Wartime Honours and Awards to Government Scientists," PRO, ADM 219/288.

11. Gannon, *Black May*, 243; Blackett, "Evan James Williams," 404–5.

12. Rosenhead, "Operational Research at the Crossroads," 9–11.

13. Roberston, "Conrad Hal Waddington," 582; Rosenhead, "Operational Research at the Crossroads," 22.

14. "Scientists' Visit to Russia Banned," *The Times*, June 15, 1945; Brown, *J. D. Bernal*, 266–67; Mott, *Life in Science*, 49.

15. Brown, *J. D. Bernal*, 485; Werskey, *Visible College*, 300–301, 318.

16. "Fundamental Research," p. 114, Blackett Papers, PB 5/1/4/10/3; Lovell, "Blackett in War and Peace," 228.

17. Nye, *Blackett*, 89; Blackett, *Fear, War, and the Bomb*, 143–44, 149–50.

18. Crook, "Case Against Area Bombing," 186n34; Zuckerman, *Apes to Warlords*, 145.

19. "An English Scientist Exposes Atomic Diplomacy," *Bulletin of Atomic Scientists*, February 1949, pp. 43, 50; Nye, *Blackett*, 3; Ash, "Orwell's List."

20. "Cosmic Ray Conference Mexico 1951," Blackett Papers, PB 1/10A.

21. Jones, *Wizard War*, 495–96, 514–16, 524.

22. Butler, "Recollections of Blackett," 154; Nye, *Blackett*, 175; "Oral History Transcript—Dr. P. M. S. Blackett," American Institute of Physics, www.aip.org/history/ohilist/4508.html.

23. Morse, *In at the Beginnings*, 291–99; Morse, "Trends in Operational Research," 159.

24. Ormerod, "Father of Operational Research," 191–92.

25. Crowther and Whiddington, *Science at War*, 199.

26. Waddington, *OR in World War II*, xv–xvi.

Bibliography

Air Ministry, United Kingdom. "The Expansion of the Royal Air Force, 1934–1939." Air Ministry and Ministry of Defence: Air Historical Branch: Narratives and Monographs, AIR 41/8. Public Record Office, Kew, U.K.

———. *The Origins and Development of Operational Research in the Royal Air Force.* London: HMSO, 1963.

Andrew, Christopher. *Her Majesty's Secret Service: The Making of the British Intelligence Community.* New York: Penguin, 1987.

Anti-Submarine Warfare Operations Research Group. *Review of Activity, 1 April 1942 to 31 August 1944.* Office of Field Service, National Defense Research Committee: Manuscript Histories and Project Summaries, 1943–1946, MLR NC-138 177. Records of the Office of Scientific Research and Development, Record Group 227. National Archives and Records Administration, College Park, Md.

Ash, Timothy Garton. "Orwell's List." *New York Review of Books,* September 25, 2003.

August, Andrew. *The British Working Class, 1832–1940.* Harlow, U.K.: Pearson, 2007.

Baker, John R. "Counter-blast to Bernalism," *New Statesman,* July 29, 1929, 174–75.

Balme, David Edward. Oral History. Imperial War Museum, London, U.K.

Bath, Alan Harris. *Tracking the Axis Enemy: The Triumph of Anglo-American Naval Intelligence.* Lawrence: University Press of Kansas, 1998.

Bell, Archibald Colquhon. *A History of the Blockade of Germany and of the Countries Associated with Her in the Great War, Austria-Hungary, Bulgaria, and Turkey, 1914–18.* London: HMSO, 1937.

Bennett, Ralph. *Behind the Battle: Intelligence in the War with Germany, 1939–1945.* Revised edition. London: Pimlico, 1999.

Benson, Robert Louis. *A History of U.S. Communications Intelligence During World War II: Policy and Administration.* United States Cryptologic History, Series IV, World War II, Vol. 8. Fort Meade, Md.: Center for Cryptologic History, National Security Agency, 1997.

Bernal, J. D. *The Social Function of Science.* London: G. Routledge & Sons, 1939.

Bialer, Uri. *Shadow of the Bomber: The Fear of Air Attack and British Policies, 1932–1939.* London: Royal Historical Society, 1980.

Blackett, P. M. S. Papers. The Royal Society Library and Archives, London.

———. "The Frustration of Science." In *The Frustration of Science*, by Daniel Hall et al. London: George Allen & Unwin, 1935.

———. "Professor E. J. Williams, F.R.S." *Nature* 156 (1945): 355–56.

———. "Evan James Williams, 1903–1945." *Obituary Notices of Fellows of the Royal Society* 5 (1947): 386–406.

———. *Fear, War, and the Bomb: Military and Political Consequences of Atomic Energy.* New York: McGraw-Hill, 1949.

———. "Operational Research." *Operational Research Quarterly* 1 (1950): 3–6.

———. "Operations Research." Review of Philip M. Morse and George E. Kimball, *Methods of Operations Research*. *Physics Today*, November 1951, 18–20.

———. "Education of an Agnostic." *Punch*, September 3, 1958, 296–98.

———. "The Rutherford Memorial Lecture, 1958." *Proceedings of the Royal Society of London. Series A, Mathematical and Physical Sciences* 251 (1959): 293–305.

———. "Charles Thomas Rees Wilson." *Biographical Memoirs of Fellows of the Royal Society* 6 (1960): 269–95.

———. *Studies of War: Nuclear and Conventional.* New York: Hill & Wang, 1962.

———. "Boy Blackett." In *Patrick Blackett: Sailor, Scientist and Socialist*, edited by Peter Hore. London: Frank Cass, 2003.

Blackett, P. M. S., and G. P. S. Occhialini. "Some Photographs of the Tracks of Penetrating Radiation." *Proceedings of the Royal Society of London. Series A, Containing Papers of a Mathematical and Physical Character* 139 (1933): 699–726.

Blair, Clay. *Hitler's U-Boat War.* Vol. 1, *The Hunters, 1939–1942.* Vol. 2, *The Hunted, 1942–1945.* New York: Random House, 1996–98.

Bowles, Edward Lindley. Papers. Manuscript Division, Library of Congress, Washington, D.C.

Brinkley, David. *Washington Goes to War.* New York: Knopf, 1988.

Brown, Andrew. "Blackett at Cambridge." In *Patrick Blackett: Sailor, Scientist and Socialist*, edited by Peter Hore. London: Frank Cass, 2003.

———. *J. D. Bernal: The Sage of Science.* New York: Oxford University Press, 2005.

Brown, David K. *Atlantic Escorts: Ships, Weapons and Tactics in World War II.* Annapolis, Md.: Naval Institute Press, 2007.

Buchheim, Lothar-Günther. *Das Boot.* 1975. Reprint. London: Cassell, 1999.

Buckley, John. *Air Power in the Age of Total War.* Bloomington: Indiana University Press, 1999.

Budiansky, Stephen. *Battle of Wits: The Complete Story of Codebreaking in World War II.* New York: Free Press, 2000.

———. *Air Power.* New York: Viking, 2004.

———. *Perilous Fight: America's Intrepid War with Britain on the High Seas, 1812–1815.* New York: Knopf, 2011.

Buell, Thomas B. *Master of Sea Power: A Biography of Fleet Admiral Ernest J. King.* Boston: Little, Brown, 1980.

Bullard, Edward. "Patrick Blackett . . . an Appreciation." *Nature* 250 (1974): 370.

Burns, James MacGregor. *Roosevelt.* Vol. 1, *The Lion and the Fox.* Vol. 2, *The Soldier of Freedom.* New York: Harcourt, Brace, 1956–1970.

Bush, Vannevar. *Pieces of the Action.* New York: Morrow, 1970.

Butler, Clifford. "Recollections of Patrick Blackett, 1945–70." *Notes and Records of the Royal Society of London* 53 (1999): 143–56.

Calder, Ritchie. *Science in Our Lives*. Revised edition. New York: Signet, 1962.

Carroll, Francis M. "The First Shot Was the Last Straw: The Sinking of the T.S.S. *Athenia* in September 1939 and British Naval Policy in the Second World War." *Diplomacy and Statecraft* 20 (2009): 403–13.

Cathcart, Brian. *The Fly in the Cathedral: How a Group of Cambridge Scientists Won the International Race to Split the Atom*. New York: Farrar, Straus & Giroux, 2005.

Christopherson, Derman, and E. C. Baughan. "Reminiscences of Operational Research in World War II by Some of Its Practitioners: II." *Journal of the Operational Research Society* 43 (1992): 569–77.

Churchill, Winston S. *My Early Life: A Roving Commission*. New York: Scribner's, 1930.

———. *The World Crisis, 1911–1918*. 1931. Reprint. New York: Free Press, 2005.

———. *The Second World War*. Vol. 1, *The Gathering Storm*. Vol. 2, *Their Finest Hour*. Vol. 3, *The Grand Alliance*. Vol. 4, *The Hinge of Fate*. Vol. 5, *Closing the Ring*. Boston: Houghton Mifflin, 1948–51.

Clark, Ronald W. *The Rise of the Boffins*. London: Phoenix House, 1962.

———. *Tizard*. Cambridge, Mass.: MIT Press, 1965.

Cohen, Eliot A., and John Gooch. *Military Misfortunes: The Anatomy of Failure in War*. New York: Free Press, 1990.

Corum, James S. *The Luftwaffe: Creating the Operational Air War, 1918–1940*. Lawrence: University Press of Kansas, 1997.

Craven, Wesley Frank, and James Lea Cate, eds. *The Army Air Forces in World War II*. Vol. 1, *Plans and Early Operations, January 1939 to August 1942*. Vol. 2, *Europe—Torch to Pointblank, August 1942 to December 1943*.Vol. 6, *Men and Planes*. Chicago: University of Chicago Press, 1948–55.

Crook, Peter. "The Case Against Area Bombing." In *Patrick Blackett: Sailor, Scientist and Socialist*, edited by Peter Hore. London: Frank Cass, 2003.

Crowther, J. G., and R. Whiddington. *Science at War*. London: HMSO, 1947.

Cunningham, W. Peyton, Denys Freeman, and Joseph F. McCloskey. "Of Radar and Operations Research: An Appreciation of A. P. Rowe (1898–1976)." *Operations Research* 32 (1984): 958–67.

Davies, Evan. "The Selborne Scheme: The Education of the Boy." In *Patrick Blackett: Sailor, Scientist and Socialist*, edited by Peter Hore. London: Frank Cass, 2003.

Dell, S. "Economics and Military Operations Research." *Review of Economics and Statistics* 42 (1960): 219–22.

Denniston, Alastair G. Papers. Churchill Archives Centre, Churchill College, Cambridge University, Cambridge, U.K.

Dönitz, Karl. *Memoirs: Ten Years and Twenty Days*. 1959. Reprint. Annapolis, Md.: Naval Institute Press, 1990.

Douhet, Giulio. *The Command of the Air*. Trans. by Dino Ferrari. 1942. Reprint. Washington, D.C.: Office of Air Force History, 1983.

Eaker, Ira C. Oral History. May 22, 1962. File Number K239.0512-627. U.S. Air Force Historical Research Agency, Maxwell Air Force Base, Ala.

Easterfield, T. E. "The Special Research Unit at the Board of Trade, 1946–1949." *Journal of the Operational Research Society* 34 (1983): 565–68.

Eastern Sea Frontier War Diary. World War II War Diaries, 1941–1945, MLR A1 353. Records of the Office of the Chief of Naval Operations, Record Group 38. National Archives and Records Administration, College Park, Md. (Available online at http://www .uboatarchive.net/ESF.htm.)

Eiduson, Bernice T. *Scientists: Their Psychological World.* New York: Basic Books, 1962.

Erskine, Ralph. "U-Boats, Homing Signals and HFDF." *Intelligence and National Security* 2 (1987): 324–30.

———. "Naval Enigma: The Breaking of Heimisch and Triton." *Intelligence and National Security* 3 (1988): 162–82.

———. "Naval Enigma: A Missing Link." *International Journal of Intelligence and Counterintelligence* 3 (1989): 493–508.

———. "Ultra and Some U.S. Navy Carrier Operations." *Cryptologia* 19 (1995): 81–96.

———. "Kriegsmarine Short Signal Systems—and How Bletchley Park Exploited Them." *Cryptologia* 23 (1999): 65–92.

———. "The Holden Agreement on Naval Sigint: The First BRUSA?" *Intelligence and National Security* 14 (1999): 187–97.

———. "The Admiralty and Cipher Machines During the Second World War: Not So Stupid After All." *Journal of Intelligence History* 2, no. 2 (Winter 2002): 49–68.

Falconer, N. "On the Size of Convoys: An Example of the Methodology of Leading Wartime OR Scientists." *Operational Research Quarterly* 27 (1976): 315–27.

Farago, Ladislas. *The Tenth Fleet.* 1962. Reprint. New York: Drum Books, 1986.

Feshbach, Herman. "Philip McCord Morse, 1903–1985." *Biographical Memoirs.* Washington, D.C.: National Academy of Sciences, 1994.

Fortun, M., and S. S. Schweber. "Scientists and the Legacy of World War II: The Case of Operations Research (OR)." *Social Studies of Science* 23 (1993): 595–642.

Fuehrer Conference on Naval Affairs, 1939–1945. London: Chatham, 2005.

Fussell, Paul. *The Great War and Modern Memory.* London: Oxford University Press, 1977.

———. *Wartime: Understanding and Behavior in the Second World War.* New York: Oxford University Press, 1989.

Gannon, Michael. *Black May: The Epic Story of the Allies' Defeat of the German U-Boats in May 1943.* New York: HarperCollins, 1998.

Gilbert, Martin. *The First World War.* New York: Henry Holt, 1994.

———. *Churchill and America.* New York: Free Press, 2005.

Gilbert, Martin, ed. *The Churchill War Papers.* Vol. 1, *At the Admiralty, September 1939–May 1940.* New York: Norton, 1993.

Goodeve, Charles. "Operational Research as Science." *Journal of the Operations Research Society of America* 1 (1953): 166–80.

Goodman, R. J., K. W. McMahan, and E. J. Carpenter. "The Battle of the Atlantic." Government Code & Cypher School History of British Signals Intelligence in World War II, Naval Vol. XVIII. National Cryptologic Museum Library, Fort Meade, Md. (Also in HW 11/31, Public Record Office, Kew, U.K.)

Graves, Robert. *Good-bye to All That: An Autobiography.* 1929. 2nd revised edition. New York: Anchor, 1958.

Graves, Robert, and Alan Hodge. *The Long Week-End: A Social History of Great Britain, 1918–1939.* 1940. Reprint. New York: Norton, 1994.

Hackmann, Willem D. "Sonar Research and Naval Warfare, 1914–1954: A Case Study of a Twentieth-Century Establishment Science." *Historical Studies in the Physical and Biological Sciences* 16 (1986): 83–110.

Haley, K. Brian. "War and Peace: The First 25 Years of OR in Great Britain." *Operations Research* 50 (2002): 82–88.

Hickman, James C., and Linda Heacox. "Actuaries in History: The Wartime Birth of Operations Research." *North American Actuarial Journal* 2, no. 4 (October 1998): 1–9.

Higgins, A. Pearce. *Defensively-Armed Merchant Ships and Submarine Warfare*. London: Stevens & Sons, 1917.

Hinsley, F. H., et al. *British Intelligence in the Second World War: Its Influence on Strategy and Operations*. 3 vols. New York: Cambridge University Press, 1979–88.

Hinsley, F. H., and Alan Stripp, eds. *Codebreakers: The Inside Story of Bletchley Park*. Oxford: Oxford University Press, 1994.

Hitch, Charles. "Economics and Military Operations Research." *Review of Economics and Statistics* 40 (1958): 199–209.

———. "A Further Comment on Economics and Military Operations Research." *Review of Economics and Statistics* 42 (1960): 222–23.

Hodgkin, Alan, et al. "Memorial Meeting for Lord Blackett, O.M., C.H., F.R.S. at the Royal Society on 31 October 1974." *Notes and Records of the Royal Society of London* 29 (1975): 135–62.

Hore, Peter. "Blackett at Sea." In *Patrick Blackett: Sailor, Scientist and Socialist*, edited by Peter Hore. London: Frank Cass, 2003.

Horvath, William J., and Martin L. Ernst. "Philip McCord Morse: A Remembrance." *Operations Research* 34 (1986): 7–9.

Hough, Richard, and Denis Richards. *The Battle of Britain*. New York: Norton, 1989.

Huxley, Julian S. "Science in War." *Nature* 146 (1940): 112–13.

Jones, R. V. *The Wizard War: British Scientific Intelligence, 1939–1945*. New York: Coward, McCann & Geoghegan, 1978.

Kahn, David. *Seizing the Enigma: The Race to Break the German U-Boat Codes, 1939–1943*. Boston: Houghton Mifflin, 1991.

Kennett, Lee. *A History of Strategic Bombing*. New York: Scribner's, 1982.

King-Hall, Stephen. *A North Sea Diary, 1914–1918*. London: Newnes, 1936.

Kirby, M. W. *Operational Research in War and Peace: The British Experience from the 1930s to 1970*. London: Imperial College Press, 2003.

Kirby, M. W., and R. Capey. "The Origins and Diffusion of Operational Research in the UK." *Journal of the Operational Research Society* 49 (1998): 307–26.

Klein, Joe. *Woody Guthrie: A Life*. New York: Ballantine, 1982.

Konvitz, Josef W. "Bombs, Cities, and Submarines: Allied Bombing of the French Ports, 1942–43." *International History Review* 14 (1992): 23–44.

Koopman, B. O. "The Theory of Search. I. Kinematic Bases." *Operations Research* 4 (1956): 324–46.

———. "The Theory of Search. II. Target Detection." *Operations Research* 4 (1956): 503–31.

———. "The Theory of Search. III. The Optimum Distribution of Searching Effort." *Operations Research* 5 (1957): 613–26.

Koopman, B. O., and Charles Hitch. "Fallacies in Operations Research." *Operations Research* 4 (1956): 422–30.

Krohn, P. L. "Solly Zuckerman, Baron Zuckerman, of Burnham Thorpe, O.M., K.C.B. 20

May 1904–1 April 1993." *Biographical Memoirs of Fellows of the Royal Society* 41 (1995): 576–98.

Larnder, Harold. "The Origin of Operational Research." *Operations Research* 32 (1984): 465–75.

Lawrence, Hal. *Tales of the North Atlantic*. Toronto: McClelland & Stewart, 1985.

Layton, Edwin T., Roger Pineau, and John Costello. *"And I Was There": Pearl Harbor and Midway—Breaking the Secrets*. New York: Morrow, 1985.

Little, John D. C. "Philip M. Morse and the Beginnings." *Operations Research* 50 (2002): 146–48.

Llewellyn-Jones, Malcolm. "A Clash of Cultures: The Case for Large Convoys." In *Patrick Blackett: Sailor, Scientist and Socialist*, edited by Peter Hore. London: Frank Cass, 2003.

Lloyd George, David. *War Memoirs of David Lloyd George*. 6 vols. Boston: Little, Brown, 1933–37.

Love, Robert William, Jr. "Ernest Joseph King." In *The Chiefs of Naval Operations*, edited by Robert William Love, Jr. Annapolis, Md.: Naval Institute Press, 1980.

Lovell, Bernard. "Patrick Maynard Stuart Blackett, Baron Blackett, of Chelsea. 18 November 1897–13 July 1974." *Biographical Memoirs of Fellows of the Royal Society* 21 (1975): 1–115.

———. "Blackett in War and Peace." *Journal of the Operational Research Society* 39 (1988): 221–33.

Mahon, A. P. "History of Hut Eight." NR 4685, Historic Cryptographic Collection, MLR A1 9032. Records of the National Security Agency, Record Group 457. National Archives, College Park, Md. (Available online at http://www.ellsbury.com/hut8/hut8-000.htm.)

Manchester, William. *The Glory and the Dream: A Narrative History of America, 1932–1972*. New York: Bantam, 1975.

———. *The Last Lion, Winston Spencer Churchill*. Vol. 1., *Visions of Glory, 1874–1932*. Vol. 2, *Alone, 1932–1940*. Boston: Little, Brown, 1983–88.

Mangel, Marc. "Applied Mathematicians and Naval Operators." *SIAM Review* 24 (1982): 289–300.

Mark, Eduard. *Aerial Interdiction in Three Wars*. Washington, D.C.: Center for Air Force History, 1994.

Mason, John T., Jr., ed. *The Atlantic War Remembered: An Oral History Collection*. Annapolis, Md.: Naval Institute Press, 1990.

Massie, Robert K. *Dreadnought: Britain, Germany, and the Coming of the Great War*. New York: Random House, 1991.

McCloskey, Joseph F. "The Beginnings of Operations Research: 1934–1941." *Operations Research* 35 (1987): 143–52.

———. "British Operational Research in World War II." *Operations Research* 35 (1987): 453–70.

———. "U.S. Operations Research in World War II." *Operations Research* 35 (1987): 910–25.

McCloskey, Joseph F., and Florence N. Trefethen, eds. *Operations Research for Management*. Baltimore: Johns Hopkins University Press, 1954.

McCue, Brian. *U-Boats in the Bay of Biscay: An Essay in Operations Analysis*. Washington, D.C.: National Defense University Press, 1990.

McMahan, K. W. "The German Navy's Use of Special Intelligence" and "German Reactions to Allied Use of Special Intelligence." Government Code & Cypher School History of British Signals Intelligence in World War II, Naval Sigint, Vol. VII. National Cryptologic Museum Library, Fort Meade, Md. (Also in HW 43/17, Public Record Office, Kew, U.K..)

Meigs, Montgomery C. *Slide Rules and Submarines: American Scientists and Subsurface Warfare in World War II*. Washington, D.C.: National Defense University Press, 1990.

Middlebrook, Martin. *Convoy*. New York: Morrow, 1976.

Miller, Nathan. *War at Sea: A Naval History of World War II*. New York: Oxford University Press, 1995.

Milner-Barry, Philip Stuart. Oral History. Imperial War Museum, London, U.K.

———. "Action This Day: The Letter from Bletchley Park Cryptanalysts to the Prime Minister, 21 October 1941." *Intelligence and National Security* 1 (1986): 272–76.

Moll, John L. "William Bradford Shockley." *Biographical Memoirs*. Washington, D.C.: National Academy of Sciences, 1995.

Monsarrat, Nicholas. *The Cruel Sea*. New York: Knopf, 1951.

Morison, Samuel Eliot. *History of United States Naval Operations in World War II*. Vol. 1, *The Battle of the Atlantic, September 1939–May 1943*. 1947. Reprint. Annapolis, Md.: Naval Institute Press, 2010.

Morris, Richard Knowles. *John P. Holland, 1841–1914: Inventor of the Modern Submarine*. 1966. Reprint. Columbia: University of South Carolina Press, 1998.

Morse, Philip M. "Trends in Operations Research." *Journal of the Operations Research Society of America* 1 (1953): 159–65.

———. "George Elbert Kimball, 1906–1967." *Biographical Memoirs*. Washington, D.C.: National Academy of Sciences, 1973.

———. *In at the Beginnings: A Physicist's Life*. Cambridge, Mass.: MIT Press, 1977.

———. "In Memoriam: Bernard Osgood Koopman, 1900–1981." *Operations Research* 30 (1982): 417–27.

———. "The Beginnings of Operations Research in the United States." *Operations Research* 34 (1986): 10–17.

Morse, Philip M., and George E. Kimball. *Methods of Operations Research*. 1946. Reprint. Mineola, N.Y.: Dover, 2003.

Morse, Philip M., et al. "Where the War Stories Began: Some Founding Fathers Reminisce." *Interfaces* 11, no. 5 (October 1981): 37–44.

Mott, Nevill. *A Life in Science*. London: Taylor & Francis, 1995.

Mowry, David P. *The Cryptology of the German Intelligence Services*. United States Cryptologic History: Series IV, World War II, Vol. 4. Office of Archives and History, National Security Agency, 1989. (Available online at http://www.nsa.gov/public_info/declass/cryptologic_histories.shtml.)

National Archives and Records Administration. Division 6 (Sub-Surface Warfare), National Defense Research Committee: General Administration Records, 1941–1946, MLR NC-138 78. Records of the Office of Scientific Research and Development, Record Group 227. College Park, Md.

———. Formerly Security-Classified Records of Edward L. Bowles Concerning Antisubmarine Warfare in the Atlantic, 1942–1945, MLR A1 117. Records of the Office of the Secretary of War, Record Group 107.

———. Historic Cryptographic Collection, Pre–World War I Through World War II, 1891–1981. Records of the National Security Agency, Record Group 457.

———. Tenth Fleet Records: Correspondence and Subject Files, 1942–1945, Anti-

Submarine Warfare Division, MLR A1 349; Records Relating to Establishment, Disestablishment, and Organization, 1942–1946, Anti-Submarine Warfare Division, Analysis and Statistical Section, Series I, MLR A1 350; Administrative Records, 1942–1947, Anti-Submarine Warfare Division, Analysis and Statistical Section, Series II, MLR A1 350. Records of the Office of the Chief of Naval Operations, Record Group 38.

National Defense Research Committee. *A Survey of Subsurface Warfare in World War II.* Summary Technical Report of Division 6, NDRC, Vol. 1. Washington, D.C.: GPO, 1946.

National Security Agency. *Battle of the Atlantic.* Vol. 1, *Allied Communications Intelligence, December 1942–May 1945* (Studies on Cryptology Series, SRH-009). Vol. 2, *U-Boat Operations—December 1942 to May 1942 Including German U-Boats and Raiders in the Indian and Pacific Oceans* (SRH-008). Vol. 3, *German Naval Communication Intelligence and Compromise of Allied Ciphers* (SRH-024). Vol. 4, *Technical Intelligence from Allied C.I., 1941–1945* (SRH-025). Vol. 6, *Appendices.* (Available online at http://www.ibiblio.org/hyperwar/ETO/Ultra/index.html.)

Nye, Mary Jo. "A Physicist in the Corridors of Power: P. M. S. Blackett's Opposition to Atomic Weapons Following the War." *Physics in Perspective* 1 (1999): 136–56.

———. *Blackett: Physics, War, and Politics in the Twentieth Century.* Cambridge, Mass.: Harvard University Press, 2004.

Oliphant, Mark. *Rutherford: Recollections of the Cambridge Days.* Amsterdam: Elsevier, 1972.

Ormerod, Richard. "The Father of Operational Research." In *Patrick Blackett: Sailor, Scientist and Socialist,* edited by Peter Hore. London: Frank Cass, 2003.

Overy, Richard. *The Battle of Britain.* New York: Norton, 2001.

Padfield, Peter. *Dönitz: The Last Führer.* New York: Harper & Row, 1984.

Parsons, John E. *Functions of "Secret Room" of Cominch Combat Intelligence Atlantic Section.* SRMN-038, U.S. Navy Records Relating to Cryptology. Records of the National Security Agency, Record Group 457. National Archives, College Park, Md.

Pile, Frederick Alfred. *Ack-ack: Britain's Defence Against Air Attack During the Second World War.* London: Harrap, 1949.

Price, Alfred. *Aircraft Versus Submarine in Two World Wars.* 1973. Reprint. Barnsley, U.K.: Pen & Sword Aviation, 2004.

Public Record Office. The National Archives of the United Kingdom. Admiralty: Correspondence and Papers, ADM 1. Kew, U.K.

———. Admiralty: Office of the First Sea Lord: Correspondence and Papers, ADM 205.

———. Admiralty: Directorate of Operational Research and Predecessors: Reports, ADM 219.

———. Admiralty: Naval Intelligence Division and Operational Intelligence Centre: Intelligence Reports and Papers, ADM 223.

———. Air Ministry and Admiralty: Coastal Command: Registered Files, AIR 15.

———. Air Ministry: Private Office Papers, AIR 19.

———. Government Code and Cypher School: World War II Official Histories, HW 11.

———. Government Code and Cypher School: Directorate: Second World War Policy Papers, HW 14.

———. Government Code and Cypher School: Histories of British Sigint, HW 43.

———. Ministry of Defence and predecessors: Air Publications and Reports, AIR 10.

———. War Cabinet and Cabinet: Memoranda, CAB 24.

————. War Cabinet and Cabinet: Minutes, CAB 65.

Ray, John. *The Night Blitz, 1940–41*. London: Cassell, 2000.

Rhodes, Richard. *The Making of the Atomic Bomb*. New York: Simon & Schuster, 1986.

Roberts, Elizabeth. *A Woman's Place: An Oral History of Working-Class Women 1890–1940*. Oxford: Basil Blackwell, 1984.

Robertson, Alan. "Conrad Hal Waddington. 8 November 1905–26 September 1975." *Biographical Memoirs of Fellows of the Royal Society* 23 (1977): 575–622.

Rosenhead, Jonathan. "Operational Research at the Crossroads: Cecil Gordon and the Development of Post-War OR." *Journal of the Operational Research Society* 40 (1989): 2–28.

————. "Swords into Ploughshares: Cecil Gordon's Role in the Post-War Transition of Operational Research to Civilian Uses." *Public Administration* 69 (1991): 481–501.

Roskill, Stephen W. *The War at Sea, 1939–1945*. 3 vols. London: HMSO, 1954–61.

Rowe, A. P. *One Story of Radar*. Cambridge, U.K.: Cambridge University Press, 1948.

Russell, Jerry C. *The Campaign Against the U-boats in World War II*. National Security Agency Studies on Cryptology Series, SRH-142. 1980. (Available online at http://www.ibiblio.org/pha/ultra/navy-1.html.)

Sassoon, Siegfried. *Siegfried's Journey, 1916–1920*. London: Faber & Faber, 1945.

Sawyer, F. L., et al. "Reminiscences of Operational Research in World War II by Some of Its Practitioners." *Journal of the Operational Research Society* 40 (1989): 115–36.

Scheer, Reinhard. *Germany's High Sea Fleet in the World War*. London: Cassell, 1920.

Schoenfeld, Max. *Stalking the U-boat: USAAF Offensive Antisubmarine Operations in World War II*. Washington, D.C.: Smithsonian Institution, 1995.

Schofield, B. B. "The Defeat of the U-Boats During World War II." *Journal of Contemporary History* 16 (1981): 119–29.

Science in War. Harmondsworth, U.K.: Penguin, 1940.

Scott, James Brown, ed. *Diplomatic Correspondence Between the United States and Germany, August 1, 1914–April 6, 1917*. New York: Oxford University Press, 1918.

Senate. United States. *Submarine Boat Holland*. Senate Doc. No. 226, 55th Cong., 2nd Sess., 1898.

————. *Submarine Torpedo Boat Holland*. Senate Doc. No. 14, 56th Cong., 1st Sess., 1899.

Shackleton, Ernest Henry. *South: The Story of Shackleton's Last Expedition, 1914–1917*. New York: Macmillan, 1920.

Shirer, William L. *The Rise and Fall of the Third Reich*. New York: Simon & Schuster, 1960.

Showell, Jak P. Mallmann. *The German Navy in World War Two: A Reference Guide to the Kriegsmarine, 1935–1945*. Annapolis, Md.: Naval Institute Press, 1979.

————. *The U-Boat Century: German Submarine Warfare, 1906–2006*. London: Chatham, 2006.

Shrader, Charles R. *History of Operations Research in the United States Army*. Vol. 1, 1942–1962. Washington, D.C.: Office of the Deputy Under Secretary of the Army for Operations Research, United States Army, 2006. (Available online at http://www.history.army.mil/html/books/hist_op_research/index.html.)

Shurkin, Joel N. *Broken Genius: The Rise and Fall of William Shockley, Creator of the Electronic Age*. London: Macmillan, 2006.

Sims, William Sowden, and Burton J. Hendrick. *The Victory at Sea*. Garden City, N.Y.: Doubleday, Page, 1920.

Slatterly, John. "Postprandial Proceedings of the Cavendish Society. I." *American Physics Teacher* 7 (1939): 179–85.

Slessor, John. *The Central Blue: The Autobiography of Sir John Slessor, Marshal of the R.A.F.* New York: Praeger, 1957.

Smith, Malcolm S. *British Air Strategy Between the Wars.* Oxford: Clarendon Press, 1984.

Smyth, Edward A., and Eugene P. Visco. "Military Operations Research Society (MORS) Oral History Project Interview of Dr. Ralph Beatty, Jr." *Military Operations Research* 9, no. 3 (2004): 45–73.

Sternhell, Charles M., and Alan M. Thorndike. *Antisubmarine Warfare in World War II.* U.S. Navy Operations Evaluation Group, Report No. 51. 1946. Reprint. Laguna Hills, Calif.: Aegean Park Press.

Stevenson, John. *British Society, 1914–45.* London: Penguin, 1984.

Stiles, Bert. *Serenade to the Big Bird.* New York: Norton, 1947.

Sueter, Murray F. *The Evolution of the Submarine Boat, Mine, and Torpedo.* Portsmouth, U.K.: Gieve's, 1907.

Syrett, David. *The Defeat of the German U-Boats: Battle of the Atlantic.* Columbia: University of South Carolina Press, 1994.

Syrett, David, ed. *The Battle of the Atlantic and Signals Intelligence: U-Boat Situations and Trends, 1941–1945.* Publications of the Navy Records Society, No. 139. Aldershot, U.K.: Ashgate, 1998.

———. *The Battle of the Atlantic and Signals Intelligence: U-Boat Tracking Papers, 1941–1947.* Publications of the Navy Records Society, No. 144. Aldershot, U.K.: Ashgate, 2002.

Tarrant, V. E. *The U-Boat Offensive, 1914–1945.* Annapolis, Md.: Naval Institute Press, 1989.

Terraine, John. *To Win a War: 1918 the Year of Victory.* 1978. Reprint. London: Cassell, 2000.

Thomas, William. "The Heuristics of War: Scientific Method and the Founders of Operations Research." *British Journal for the History of Science* 10 (2007): 251–74.

Tidman, Keith R. *The Operations Evaluation Group: A History of Naval Operations Analysis.* Annapolis, Md.: Naval Institute Press, 1984.

Tizard, Henry T. Papers. Imperial War Museum, London, U.K.

van der Vat, Dan. *The Atlantic Campaign.* 1988. Reprint. Edinburgh, U.K.: Birlinn, 2001.

Waddington, C. H. *O.R. in World War 2: Operational Research Against the U-Boats.* London: Elek, 1973.

Waddington, C. H., Charles Goodeve, and Rolfe Tomlinson. "Appreciation: Lord Blackett." *Operational Research Quarterly* 25, no. 4 (December 1974): i–viii.

Webster, Charles, and Noble Frankland. *The Strategic Air Offensive Against Germany, 1939–45.* 4 vols. London: HMSO, 1961.

Wells, H. G. *Experiment in Autobiography.* New York: Macmillan, 1934.

Werrell, Kenneth P. "The Strategic Bombing of Germany in World War II: Costs and Accomplishments." *Journal of American History* 72 (1986): 702–13.

Werskey, Gary. *The Visible College: The Collective Biography of British Scientific Socialists of the 1930s.* New York: Holt, Rinehart & Winston, 1978.

Whitehead, David. "Cobra and Other Bombes." *Cryptologia* 20 (1996): 289–307.

Whitman, Edward C. "John P. Holland—Father of the Modern Submarine." *Undersea Warfare,* Summer 2003.

Williams, E. C. "Reflections on Operational Research." *Journal of the Operations Research Society of America* 2 (1954): 441–43.

———. "The Origin of the Term 'Operational Research' and the Early Development of the Military Work." *Journal of the Operational Research Society* 19 (1968): 111–13.

Williamson, Rajkumari, ed. *The Making of Physicists.* Bristol, U.K.: Adam Hilger, 1987.

Wohl, Robert. *The Generation of 1914.* Cambridge, Mass.: Harvard University Press, 1979.

Woolf, Leonard. *Downhill All the Way: An Autobiography of the Years 1919 to 1939.* 1967. Reprint. New York: Harcourt, Brace, Jovanovich, 1975.

———. *The Journey Not the Arrival Matters: An Autobiography of the Years 1939 to 1969.* 1969. Reprint. New York: Harcourt, Brace, Jovanovich, 1975.

Zimmerman, David. "Preparations for War." In *Patrick Blackett: Sailor, Scientist and Socialist,* edited by Peter Hore. London: Frank Cass, 2003.

———. "The Society for the Protection of Science and Learning and the Politicization of British Science in the 1930s." *Minerva* 44 (2006): 25–45.

Zuckerman, Lord. *From Apes to Warlords: The Autobiography (1904–1946) of Solly Zuckerman.* London: Hamilton, 1978.

———. *Six Men Out of the Ordinary.* London: Peter Owen, 1992.

Index

Illustration Credits

ALSO BY STEPHEN BUDIANSKY

PERILOUS FIGHT

*America's Intrepid War with Britain
on the High Seas, 1812–1815*

In *Perilous Fight*, Stephen Budiansky tells the rousing story of
the U.S. Navy during the War of 1812, when an upstart American fleet fought off the legendary Royal Navy and established
America as a world power for the first time. Through vivid
re-creations of riveting and dramatic encounters at sea, Budiansky shows how this underdog coterie of seamen and their
visionary secretary of the navy combined bravery and strategic
brilliance to defeat the British, who had dominated the seas for
more than two centuries. A gripping and essential history, this
is the military and political story of how the U.S. Navy became
a permanent and essential part of the nation's defense.

History